Sand

SAND

Raymond Siever

SCIENTIFIC
AMERICAN
LIBRARY

A division of HPHLP
New York

Library of Congress Cataloging-in-Publication Data

Siever, Raymond.
  Sand/Raymond Siever.
    p. cm.
  Bibliography: p.
  Includes index.
  ISBN 0-7167-5021-X (Freeman)
  1. Sand. I. Title.
QE471.2.S545 1988   87-34181
553.6'22—dc19   CIP

Printed in the United States of America

Scientific American Library
A division of HPHLP

Distributed by W. H. Freeman and Company,
41 Madison Avenue, New York, New York 10010 and
20 Beaumont Street, Oxford OX1 2NQ, England

1 2 3 4 5 6 7 8 9 0 KP 6 5 4 3 2 1 0 8 9 8

This book is number 24 of a series.

To Francis and Paul,
companions in sandstone geology,
and to Doris, Larry, and Michael,
companions on sandy beaches

Contents

Preface

Most geologists have a favorite kind of rock. Mine is sandstone. Other rocks may be more beautiful, varied in texture, or colorful, but, somehow, sand appeals to me. Regardless of my appreciation for sandstone, however, there is reason enough for writing a book like this one for a general audience. Sand and sandstone contain more information than any other sediment, and that information is accessible to us through simple physical and chemical tests. The study of sand pays off when we want to know about the geological processes that control the shape of the Earth's surface or when we attempt to reconstruct the geology of times past when a sandstone was deposited. It pays off practically as well, as those in the oil and gas industry know, because oil deposits lie in sandstone.

Why should the general reader learn about sand? First, it reveals much about the geology of the Earth's surface, the thin interface between solid Earth and the gaseous atmosphere, where we spend our lives. Learning about sand means learning about mountains, rivers, deserts, glaciers, and ocean floors. The same processes that form, transport, and deposit sand grains also carve the land surface into its familiar shapes. The landscape of the continents is the complex set of forms taken by rocks from the Earth's interior through exposure to the wind, water, and ice of the atmosphere. Sand is a tracer of all the movements, fast and slow, of the Earth's materials as they change their surroundings and produce the scenery of our lives.

Second, as records of the past, sandstones reveal the history of our planet. Specific properties of sandstones help us to construct models of the mountains and oceans of the past, by which we can trace the changing geography of the continents as they drifted, witness the birth and death of great mountain chains, and picture the varied environments in which life evolved. Geologists say that the present is the key to the past. It is also valid to see the present as the product of our past, for we cannot understand the present topography of the mountains of Pennsylvania without knowing something of their long and varied history.

Curiosity about sand develops from our wandering along beaches, from our seeing mountains of sand in dune fields, and from our taking trips to such scenic areas as the southwestern United States. In a way, I have written this book for friends who have asked me questions and for all those who have occasionally wondered about this stuff that runs through our fingers and that has given us so many metaphors, from the sandman who puts children to sleep to the hourglass of time.

In telling this story I have tried to give a glimpse of the people who have enlarged our understanding of this basic material, the geologists who found out how sand fits in with the rest of the machinery of the Earth. One part of a scientist's love for science is the interplay with people that is so much a part of the process of enunciating new theories. So I have allowed the human side to take its place with the development of ideas, enough to give a suggestion of the complex people who have investigated sand and sandstone.

Two of those people are central to this book: Francis Pettijohn and Paul Potter. In a way, this book began when in the early 1960s Francis called Paul and me, two of his former students, to work with him in teaching a short course in sandstone to a group of Canadian oil geologists. The three of us became a team and found great enjoyment in the work and in each other's company, all in the shadow of the Canadian Rockies. In the course of preparing the syllabus and giving an intensive week of lectures for that course, we realized how much of sedimentary geology—indeed, all of geology—could be encompassed by the study of one rock type. After giving the course another time, we decided to expand our weighty syllabus into a book. A year ago, the second edition of that text appeared, just about the time that the editors of the Scientific American Library talked to me about doing a more general book for this series. Linda Chaput and I discussed several choices, but it was Gerard Piel who urged me to do *Sand,* and I am glad that he did.

This is a different book, written for a very different purpose, than the advanced text that came before. To show the ways in which sand illustrates the geology of the Earth's surface, this book covers a broader span of the science of the Earth. At the same time, the lineage from the earlier book is clear. In particular, I have tried to give some of the flavor of current research in one of the most active fields of geology and have sought to explain some of the advanced ideas that would normally be the province of graduate students writing Ph.D. theses.

Many people have contributed to putting this book together. Francis and Paul, of course, had most to do with it, although their contribution is indirect, and they should not be considered responsible for any deficiencies. Linda Chaput, Gerard Piel, and Jeremiah Lyons counseled and kept me at it through various episodes of turmoil. Georgia Lee Hadler ably directed the editing process, Lisa Douglis created the elegant design, and Travis Amos was enormously helpful in finding photographs that illustrate the many different faces of sand. Doris read, asked questions, and, as always, was an incomparable intellectual sounding board.

Cambridge, Massachusetts
January 1988

Sand

CHAPTER ONE

What we can learn from sand

To see a World in a Grain of Sand
And a Heaven in a Wild Flower
Hold Infinity in the palm of your hand
And Eternity in an hour

Better than any dry description, William Blake's vision of a grain of sand as a microcosm says what this book is about. A single sand grain—an irregularly shaped fragment of rock—is the mute record of former mountains, rivers, and deserts, and of millions of years of the Earth's upheavals and quiescence. To make a grain tell us its history we tear it apart bit by bit to find out its crystal structure, its chemical composition, its radioactive age, its external shape, and its internal strain. Yet we cannot tell all we want to know of a sand grain's origin from its composition alone, any more than we can deduce political history from human physiology. The context of the state of the world's continents and oceans at a particular time is the background. That grain was produced by forces that made the rock it was eroded from, by the Earth's surface environment that eroded it from its parent and carried it to a resting place, and by the internal deformation of the Earth's crust that buried it.

In this book we will look with new eyes at a perfectly ordinary material. At the beach, we lie on sand and it gets in our shoes. We know it as a construction material, as the raw material for glass making, and as an abrasive. For most of us, sand is in the background, and we rarely focus on it. Children building sand castles

Navajo sandstone, Vermilion Cliffs, Arizona.

pay closer attention to it, manipulating moist sand easily but watching the forms collapse to dry little mounds as the water evaporates. We seldom look closely enough to see the many individual shapes, sizes, and colors of sand grains on a beach or compare those from different beaches. Some sands are pepper and salt; others, all white. Light sands may consist of almost perfectly spherical, smooth grains of a single mineral, quartz (SiO_2), one of the simplest chemical compounds of the Earth's crust. Other sands are dark grey mixtures of volcanic rock fragments and complex silicates, the chemical building blocks of much of the Earth. Where we find these sands today tells us part of the story of how the Earth's surface dynamics produced and transported sand and all the other debris of erosion over the globe. Sands can tell us how weathering and erosion work, how rivers and winds take bits and pieces of the planet around the world, and how currents flow at the bottom of the ocean.

Sand not only tells us about the dynamics of the surface of the planet today, it is a lens for reading pages of Earth history. Sandstones—ancient rocks made of sand—were formed millions of years ago in deserts, on beaches, along rivers, or under ocean waves. Volcanic grains in sandstones tell us of the former presence of volcanic oceanic islands and sand grains of some fossil shells suggest past coral reefs.

The carving of sandstones by rivers and wind gives us memorable scenery as it exposes features diagnostic of its origin. In Utah's Zion Canyon the stratification that gives texture to the great cliffs of the Navajo Sandstone indicates that these sands were formed as large dunes on a desert. We can appreciate the stunning design of nature as an esthetic at the same time that we recognize the mechanistic explanation for that design. For the sandstone geologist, the sands of today or the sandstones of the past are archives of the Earth's surface and of the forces that have shaped it since the planet was formed. While reading them we get to see some of the world's most exhilarating scenery.

HOW WE LOOK AT SAND

An analysis of the properties of sand can be as great a trial for the inexperienced geology student as a full medical history is for the medical student; both must sift through all the symptoms to find

Properties of sand and sandstone

Property	Method	Interpretation
Mineralogy	Polarizing microscopy, X-ray diffraction	Source rocks modified by weathering, transport
Chemistry	Chemical analysis	Source rocks modified by environment, diagenesis
Size	Sieveing, settling–velocity, microscopy	Transport mechanics operating on source materials
Shape and roundness	Microscopy	Abrasion during transport operating on source materials
Surface texture	Scanning electron microscopy	Weathering, environment, and diagenesis
Porosity, permeability, density	Physical methods	Environment and diagenesis operating on source materials
Cross-bedding, ripples, other sedimentary structures	Compass and angle in the field	Paleocurrents, environment
Sand body size and shape	Outcrop analysis, drill holes	Environment

Note: Properties listed are the major characteristics described and measured for the interpretation of source rocks, weathering, transport, environment, and diagenesis, the chief topics of this book. A host of other properties, such as electrical, magnetic, and mechanical, are measured for special purposes.

the pattern that triggers a diagnosis. Like medical symptoms, some properties of sand are diagnostic, others are not; we search through a long list of physical and chemical properties before we can begin to determine a sand's origins. That list includes such properties of individual grains as their mineralogy, commonly expressed as the relative proportions of the major constituent minerals. Mineralogy refers to the combination of chemical elements in specific crystalline forms of chemical compounds that make up the sand. To a geologist, it is important to know that the abundant silicon and oxygen in many sands are present not only as silicon dioxide but as the mineral quartz. A mineral-blind geologist would be like a color-blind painter: able to do some things

Sand grains *(left)*, including many volcanic rock fragments, and sandstone cliffs *(right)* eroding to provide some of the sand on the beach at Cape Arago, Oregon.

well but still powerfully handicapped. Nonmineralogical properties, such as grain size, are given as an average of many grains. Indeed, the size defines particulate material. Following centuries of common usage, geologists formally define sand as a particulate material with grain diameters between $\frac{1}{16}$ (0.0625) and 2 mm (millimeters). Particles larger than 2 mm are called granules, pebbles, or cobbles; those smaller than $\frac{1}{16}$ mm, silt grains. Because they may contain some silt, pebbles, and clay [grains smaller than $\frac{1}{256}$ (0.0039) mm], most sands are not bound by classifications. The sizes were chosen as powers of two to give a geometric scale with a suitable number of divisions.

Properties of the bulk, such as density or porosity, depend on the way in which a few thousand grains are packed together. A thousand grains of sand, each 0.5 mm in diameter, with an average amount (about 40 percent) of pore space between them would make a cube a little less than 1 cm (centimeter) on a side. Some sands are packed more tightly and take up less space. Sandstone is a rock, typically denser than sand by virtue of its mineral ce-

Ripples in sand form in response to the current action of wind or water. Their size and spacing increase as the current velocity increases.

ments, chemical precipitates added to the sand during millions of years of burial beneath other sediments.

Life would be simple for the geologist if sand were homogeneous and the properties of a 0.5-cm cube were sufficient to characterize the sand of a beach that is 100 m (meters) wide and 50 km (kilometers) long. We could pick up a few samples, take them to our laboratory for analysis, and relax. As it turns out, inhomogeneities give us much of the information we need in order to decipher the origin of such a body of sand. Walking over the beach, we might notice large grains of sand mixed with small pebbles just on the shoreline, where the waves break energetically. Looking away from the shore we might be struck with the much smaller sizes of sand higher up on the beach, where low sand dunes border the strand, or shore area. We start to differentiate regions of sand with different sets of properties.

The sand of dunes there shows rippled surfaces, an example of a larger-scale property of the accumulation of sand. Explaining the ripples' form and size by integrating over a large area the spatial distribution of the constituent grains, each with its own size, shape, and density, would be as difficult—and inappropriate—as attempting to characterize the motion of a wave breaking on the shoreline by working upward in scale from the statistical mechanics of the constituent water molecules. Instead of summing the discrete grains to describe a ripple, we treat the sand body as a continuum: we describe it the way we do any wave form, in terms of wavelength and amplitude.

Our perception of sand keeps jiggling between its large-scale and small-scale properties, between its continua of ripples and its particle constituents. On a moderately windy day, we can directly observe the movement of individual sand grains on the larger rippled surface. If we clock the correlation between the wind speed, the size of the grains transported, the trajectories along which the grains jump on the surface, and the form and size of the ripples, we can see how all these small motions add up to the larger wave form. If we wade into shallow water, we can see a different sort of ripple driven by waves in the surf zone. Here, even before we explore a concept such as transport mechanics, we can get the germ of an idea for interpreting the past current that deposited the sand. If we measure the ripple wave form and the grain size of an ancient sandstone, we might be able to discriminate between

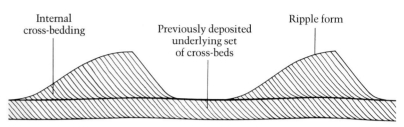

Left: Ripplemarked sandstone from an ancient lakeshore. *Right:* Internal structure of ripples showing cross-bedding.

wind-driven and wave-driven forms. If we find both types, we might infer that the two types of current interacted in a pattern characteristic of a beach. Using simple observations like these, we can read past events from relic structures preserved in the rock.

Ripples and other sedimentary structures—most are the varied forms of stratification—are preserved in many sandstones. We can see them clearly if the rock breaks along a bedding plane, revealing the form on the surface; we more commonly see cross sections of sandstones in outcrops such as canyon walls or road cuts. If we make a vertical cut through a horizontally bedded sandstone, we get a third dimension to the flat or rippled surface of the bed. We can get the same view of a beach by digging a trench. That third dimension, which is part of the ripple's structure, displays the cross-bedding: sets of beds deposited at an angle to the horizontal bedding plane. The cross-bedding gives us a more complete description of the ripple and a better basis for reconstructing the dynamics that shaped it.

We increase the scale of all three dimensions by constructing regional pictures of the full thickness of the formation from measurements at many individual outcrops and drill holes. Synthesizing all these tells us how the sand accumulated in thickness over a large area for a long period of time. For example, the total thickness of the sand body, perhaps as much as 50 m, may represent thousands of years of sedimentation. The shape of the entire sand body, long and narrow with a lenslike cross section, is also typical of a beach sand. On an even larger scale, we can map the distribution of the sand body of the beach in relation to the shallow marine sands and muds offshore and to a belt of coastal dunes in back of the shore. Finally, we can map larger regions or even major portions of continents to see the geographic patterns of rivers, desert dune fields, deltas, beaches, and continental

Cross-bedded sandstone at Zion Canyon, Utah.

shelves. That mapping fascinates us, for it gives us a look at the Earth of the past. And as we consider the cumulative history of a growing continent, we discover something of the kinds and rates of stately changes that govern the surface of the planet.

THE TOOLS WE USE

"Use your eyes before you bash the rocks," is an admonition I constantly make to geology students armed with hammers. The naked eye is a wonderful instrument for studying sandstones in the field: the eyes detect subtle differences in texture and composition brought out by weathering, much in the way photographic developers can bring out the delicate latent image on the negative. Cross-bedding is almost always easier to see on an old, exposed surface than on a fresh face cracked open by a hammer. But the very process that shows up delicate surface structures degrades the original composition of the rock and hinders our recognition of the original minerals. The chemical attacks of water, air, and organisms on the rock may oxidize and dissolve its less stable

Photomicrographs under polarized light (crossed nicol prisms). *Top:* A pure quartz sandstone consisting of rounded detrital grains cemented by quartz precipitated between the grains during post-depositional burial. *Bottom:* In a feldspathic sandstone the feldspar grains are distinguished by their light-and-dark layering.

constituent minerals. Pyroxenes, or iron silicates, break down quickly, and their decay products include the iron oxide that gives the brownish or reddish color you see as surface stains on so many sandstones. To see the original composition, the geologist hammers out a fresh broken surface that has never seen rain or bacteria.

The untutored eye may not perceive much on the broken surface: the irregularities of color, the luster, and perhaps the outlines of some larger grains seem to give little information. With practice and some knowledge of mineralogy we might make out the more translucent, lighter grains of quartz and the duller, brownish fragments of feldspar. Flakes of bright mica glint in the sun, and a few dark grains pepper the light background. But few geologists would want to look without a hand lens that magnifies eight or ten times over the naked eye. The hand lens is enough to identify most of the major minerals in all but the finest of sands, but the information is qualitative and crude. We want to know the exact kind of feldspar or mica and its composition. For that we need a microscope.

In 1958, three decades after William Nicol invented his calcite crystal polarizing prism, now called a nicol, Henry Clifton Sorby, an English geologist equally at home in the laboratory or in the field, laid out the principles of polarizing microscopy for minerals. Using thin sections—slices of rock ground to transparency at a thickness of about 0.03 mm—he showed how the optics of polarized light could be used both to identify minerals and to study the indescribable variety of textures by which mineral crystals and grains fit together to make a rock. Although Sorby had made wide-ranging investigations in geology and other sciences, he took special interest in sandstone and had measured cross-bedding in outcrops along English and Scottish rivers. He soon realized that he could infer the currents of ancient rivers from those measurements. Using the petrographic microscope, a compound microscope fitted with one polarizing prism below the thin section and another above it, he could determine the relative proportions of quartz and of different types of feldspar, mica, and other mineral grains. Even more revealing were the textures: some grains were rounded and stacked together something like closely packed cannonballs; other grains of widely different sizes were angular, and bound with a fine-grained matrix; and many

 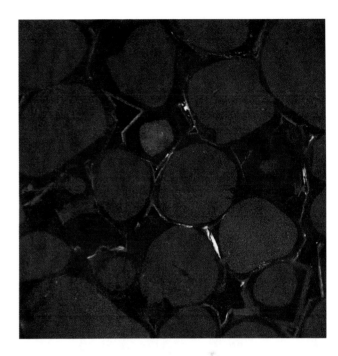

Left: Photomicrograph of quartz sandstone in polarized light. Though there is abundant quartz cement, it cannot easily be distinguished from detrital grains of quartz in this kind of illumination. *Right:* The same field under cathodo-luminescence, light emitted by the material under bombardment by electrons. In this dim light, the cementing quartz is black (non-luminescent) whereas the detrital grains are purple and violet.

others showed irregular crystals of quartz or calcite, which could not have been the original grains carried along by a stream but must have been chemically precipitated in the pore space between the grains.

Since Sorby's time, optical microscopes have been the standard instrument for mineralogical and textural analyses of rocks. Although we now have far more sophisticated tools, for the sandstone geologist, the petrographic microscope remains the equivalent of the physician's stethoscope. This tool is basic for mineralogical analysis, which is necessary to determine which rocks eroded to yield the mineral grains. It is indispensible for textural analysis, which helps us to understand the postdepositional processes that cement sand grains together to form dense rock.

Other, newer instruments have rapidly moved into widespread use, and they too are becoming indispensible. The higher magnifications of the scanning electron microscope (SEM) enables us to see important details of grain surfaces, crystal faces, and pore spaces. The morphologies seen under the SEM have changed ideas

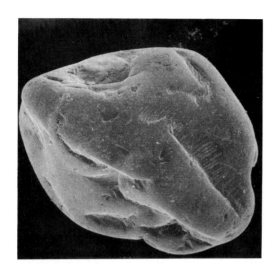

Scanning electron micrograph of a rounded and well-worn quartz sand grain from the upper part of a beach derived from the erosion of glacial deposits. Magnification × 65.

that took hold in the period when even the highest-power microscope could not resolve geometries on the scale of fractions of a micrometer. When the SEM is linked to an energy-dispersive X-ray spectrometer, we get an approximate chemical analysis of a minute area and determine how compositions vary from grain to grain. If we need more precise chemical analysis we move to the electron microprobe, dubbed familiarity "the probe" in most geological laboratories. Using more conventional wavelength-analyzed X rays, the probe can give us detailed and accurate chemical analyses over areas as small as a few square micrometers. Using the probe we can determine how the proportions of calcium and sodium in a feldspar change from the interior to the edge of a grain, which reflects the difference in geochemical environments that shaped the core and outer envelope of the grain.

Geologists long to see more of the Earth than the rocks that appear at its surface, exposed by the random work of erosion or the engineering of roads. Many centuries ago, mines offered a few restricted exposures: the mining geologist could see only the relatively shallow shafts and mine workings that covered small areas. The first derrick that drilled for oil in Titusville, Pennsylvania, in 1859, signaled the real beginning of modern subsurface geology. As oil derricks multiplied throughout the world, new information from the subsurface flooded geology and became a special boon for sedimentologists: ever since those early days, the prime target of the drillers has been the "sandrock" whose pores hold the oil.

The drilling technology determined the amount of geological information we got. The early wells were drilled by alternately hoisting and dropping a heavily weighted steel tool on a cable. Every few feet the driller would stop to bail out—the bottom of the hole was always filled with water seeping in from the moisture-saturated rocks. The pieces of crushed rock that had been penetrated in the last few feet came up in the bailer. Fragments were large enough to be identified with a hand lens or the low-power binocular stereoscopic microscope that gradually came into wide use. As the drill went deeper, the geologist constructed a log of all the formations penetrated, a crude but extraordinarily useful guide to what lay beneath.

As the technology changed, so did the information. A half-century ago the old cable-tool derricks were replaced by the mod-

Signal Hill oil field in southern California, about 1935, drilled by rotary drills.

ern rotary drill, which drilled holes several orders of magnitude faster and deeper. The rotary table at the top drove a long string of sectional pipe with a rotating toothed bit at the bottom that tore apart the rock under part of the weight of the entire length of pipe suspended from the top of the derrick. Water under high pressure flowed down the inside of the drilling pipe, out, and up through the annulus, or ring space, between pipe and rock walls, carrying with it the fine fragments of rock to the surface. The geologist now had a continuous record of samples.

At the same time, geophysical measurements took the stage, and a totally new kind of information became available. Electric

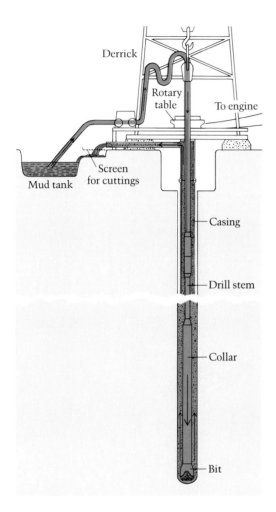

Left: The modern rotary drill employs high-pressure drilling "mud" to carry cuttings to the surface. *Right:* Workers adjusting rotary-table components.

logging, or recording, of drill holes started in the 1920s in France, spread quickly to Venezuela, the USSR, and Rumania, and by the mid-1930s rapidly came to dominate well logging in the United States. Electric well logs are produced by lowering an array of electrodes into the drill hole and measuring a variety of electrical properties of the surrounding rocks. From the different curves that come from various configurations of electrodes in the array, suitably calibrated, the geologist can infer the rock type, the permeability, and the type of fluid in the rock, whether oil, water, or

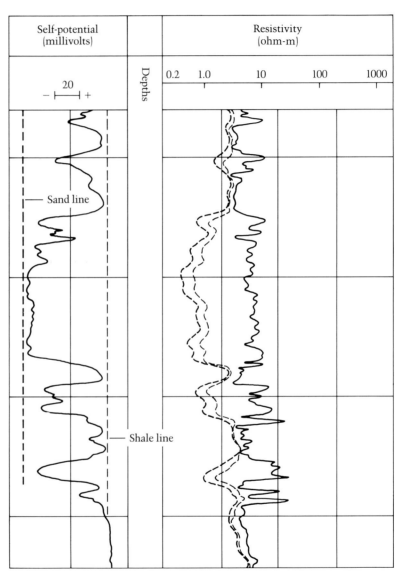

Left: The drill bit is comprised of tungsten carbide "teeth" embedded in three roller cones that are free to turn as the bit rotates. *Right:* An electric well log. First, in response to an applied current, the resistivity log measures the variation of potential between the ground surface and an electrode in the drill hole. Second, the self-potential curve measures the naturally arising electrochemical and electrokinetic potential between two electrodes in the drill hole.

gas. An added, vital dividend was the precise measurement of depth in the hole for each wiggle of the curve. Geologists, geophysicists, geochemists, and engineers have collaborated to produce such a sophisticated science and technology. It allows us to map the shape of a particularly permeable sand body, one in which the pore space can readily transmit flows of liquid; to estimate the amount of oil in the pore space of the sandstone and how much of it might be producible; to calculate the response of a rock section to seismic waves; and much more.

THE USES OF SAND

Sand is so widespread over the face of the Earth that almost all cultures have used it for construction, for abrasives, and for hundreds of other practical purposes. It is added to plaster, concrete, structural clay products, and asphalt. Sand made chiefly of quartz is ideal for these purposes because of its bulk, its relative chemical inertness, and its hardness. Other constituents of sand may be far more valuable than the sand itself: for example, the pores of sandstone carry much of the world's oil and gas. Gold as well as significant supplies of uranium and other metals are constituents of sandstones. Much of our drinking water comes from water-filled sandstones underground.

The small number of sandstones made up of virtually pure quartz—some are more than 99 percent pure—supply the world's silicon, which we use for everything from silicon chips for microprocessors to silicon chemicals, such as silicon greases and sprays, to glass. Not suprisingly, famous glassworks were built near pure quartz sand deposits. As we will see, these pure gleaming sands also have special geological significance.

We not only use sand recreationally, but expect to find generous supplies of it at any shore resort. Vacationers' demands for sandy beaches mean that the resort industry must truck in huge quantities of sand to restore beaches that have been eroded as a result of poorly thought out engineering on the shoreline. Some beaches have been made where none existed before. With this, we return full circle to the dynamics of sand movement—its source, supply, and removal—that lie at the heart of the distribution of sand now and in the past. Sand grains are produced by one set of processes, transported from their formation site to an area of deposition by another group of processes, and buried and transformed by yet another set of geological dynamics. Of these dynamics, the

Pure quartz sand from Belize, Central America. The grains are angular, showing little evidence of rounding during transport.

production of sand at its source is fundamental. Where and how are the sand grains first made? In the next chapter we discuss the birth of sand grains as their parent rocks disintegrate by weathering and erosion.

CHAPTER TWO

Where does sand come from?

S and grains are polyglot: they speak to us in terms of mineral and chemical composition, grain texture, size, geologic age, and other properties too numerous to describe. Geologists, like spies sifting through cans of garbage, looking for revealing bits of information, search for the properties that are clues to Earth's patterns of behavior now and in the past. Like garbage in a city dump, sand grains are mixed from many sources. And like the macerated refuse buried in landfills, sands may be decayed and altered almost beyond recognition. Despite these limitations, we can identify and reconstruct much of their remarkable story.

A simple tally of the different kinds of materials composing the sand is the best place to start. Silicate minerals are by far the most abundant, with quartz the most common, followed by feldspar. Silicate sands are so common because silicates are the dominant minerals in the Earth's crust. For every silicate we see as a sand grain, we can identify the corresponding silicate—even its precise chemical composition—in the igneous, sedimentary, or metamorphic rock that was its parent. We can do the same with much less abundant iron and other metal oxides such as hematite (Fe_2O_3) and rutile (TiO_2). These and many other minerals found in small amounts in sand have one chemical property in common: they are all extremely insoluble in water. Their insolubility makes them survive the rains, rivers, and seawaters that carry them to their resting place, the geological environment in which they are deposited. But insolubility isn't a characteristic of all sands.

Millions of midwinter vacationers in the Caribbean or the Bahamas lie on beaches of calcite and aragonite, two different crystalline forms of $CaCO_3$. Hence we speak of calcium-carbonate, or more simply, carbonate sands. Even the naked eye can see that most of these sand grains are fragments of shells. Like the sili-

White Sands National Monument, New Mexico.

Left: Calcium carbonate sand, Cooper's Bay, New Zealand. Many of the grains are shell fragments.
Right: Calcium carbonate sand, composed mainly of oolites, cemented slightly by calcium carbonate precipitated between the grains.

cates, they can be matched with their parents: the conches, whelks, and periwinkles that live in the shallow warm seawaters off the beach. Others are spherical grains made up of concentric precipitates of carbonate: these oolites are formed in shallow agitated waters at the edges of marine banks. The source materials for carbonate sands are produced in the sedimentary environment, giving us clues to the nature of that environment and especially to the shelled organisms that lived there. Ancient carbonate sands tell us of coral reefs, shallow carbonate banks, and beaches. Because carbonate is much more soluble than silicate or oxide in the fresh waters of the continents, carbonate sands tell us nothing of such processes as streams eroding mountains over the slow course of geologic time. Only under rare conditions of extremely rapid erosion will the carbonate sand grains of preexisting limestones survive rain and river water. But seawater, especially in the tropics, is supersaturated with carbonate, and there it does not dissolve.

Even stranger minerals can form sands. The white dunes of White Sands, New Mexico, are made of crystals of gypsum $(CaSO_4 \cdot 2H_2O)$, a moderately soluble mineral that can survive only in that arid region. Here, too, the source is local. In a valley flat, gypsum sand grains are formed by the dessicating salty brines that derive from the dissolution of older gypsum deposits. Although the gypsum is chemically derived from the older gypsum

Gypsum sand dunes of White Sands, New Mexico. *Left:* Rippled surface of dune. *Right:* Microscopic view showing the roughly rectangular shapes typical of gypsum grains and the rounding caused by abrasion during wind transport.

deposits, the grains are not the same crystals and do not "remember" anything from the older past.

Most sand grains are made of quartz and other silicates and do have some memory of their birthplace. At first glance, sand seems to be a ground-up product of eroding parent rocks. Although rivers that drain from granitic terranes carry grains of all the constituent minerals of the granite, the minerals are not in the same relative abundance that we find in the parent rock. Almost universally, we see, for example, less feldspar relative to quartz in the river sand than in the original granite. The filter that changes the relative composition is weathering, the chemical decay of unstable or soluble minerals.

Weathering we all know something about. You do not have to be a geologist (though it helps) to have a weather-beaten face. For preservation, we paint our wooden houses and wax our steel cars. If we become aware of less immediate concerns, we may notice deteriorating monuments of stone, or the gentle rounding the years impose on the angular, jagged rocks of road cuts. With or without acid rain, and even in the country, far from city fumes, all rocks weather. The rivers of Surinam give an extreme example of what weathering will do as a filter changing the relative abundance of minerals. The erosion of deeply weathered, ancient granite lowlands in that moist, tropical part of South America gives a pure-quartz river sand, with no trace of the many other minerals

The texture of feldspar in a granite. Crystals freed from this rock by weathering are angular and rhomboidal, inheriting their shape from the parent rock.

that make up the granite. The corroded, pitted quartz shows signs of etching by the weathering solutions that completely dissolved less-resistant minerals. Much of the collective memory of the granite, the parent rock, is thus erased. At the opposite extreme are rivers that drain high granite mountains in Alaska and carry dissaggregated minerals of the parent rock in almost the same proportions as the original granite. Here the memory is complete.

THEORIES OF CHEMICAL WEATHERING: THE DECAY OF FELDSPAR

This kind of match and mismatch of sand and parent-rock minerals that geologists have observed in the last century has led to modern theories linking weathering, climate, and topography. Weathering is most intense in hot, humid lowlands and least intense in cold, dry mountains. As rocks are broken into boulders, pebbles, and sand grains, weathering alters some silicates to clays, dissolves others completely, and with the help of the biological world, produces soil. How does this happen?

As much as we can learn of the overall operation of this process from the field, it remained for experimental sedimentary geochemists in the last few decades to demonstrate the mechanisms of weathering transformations. These chemical-weathering mechanisms were worked out by intensive study of one mineral group, the feldspars. Feldspar, the most abundant mineral of the Earth's crust, is the pink-to-gray mineral that appears so prominently in granites and many other igneous and metamorphic rocks. Its three pure varieties are orthoclase, potassium feldspar ($KAlSi_3O_8$), and the two plagioclase feldspars: albite, or sodium feldspar ($NaAlSi_3O_8$), and anorthite, or calcium feldspar ($CaAl_2Si_2O_8$). Most feldspars are crystalline solutions of variable composition. The alkali feldspars are mixtures of sodium (Na) and potassium (K), the plagioclases of sodium and calcium (Ca). The kinds of feldspars found in an igneous or metamorphic rock are determined by the relative abundance of K, Na, and Ca in the melt or precursor solid from which the feldspar formed. Feldspar compositions are also strongly affected by the temperatures and pressures under which they crystallized. These variables of

The chemical composition of the feldspars represented as a triangle whose apexes are the pure potassium (KAlSi₃O₈), sodium (NaAlSi₃O₈), and calcium (CaAlSi₂O₈) forms. Intermediate crystalline solutions are represented as percentages of a component, albite. Chemically stable feldspars are the alkali series, containing only small amounts of calcium, and the plagioclase series, with only small amounts of potassium. Compositions are commonly determined by polarizing microscopy, by X-ray diffraction, and by electron probe.

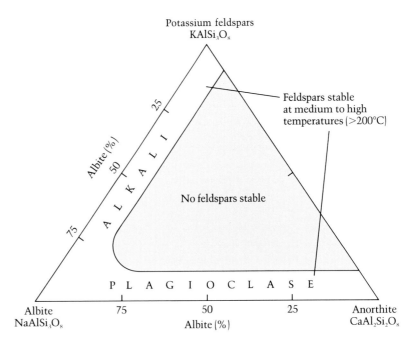

composition make it possible to analyze feldspar sand grains and read from them a great deal about the origins of their parent rocks. Of parent rocks and their relation to the mountains of the past, this chapter says more later.

The behavior of feldspars when immersed in water reflects their structure and composition. All are three-dimensional frameworks of strongly bonded sodium, potassium, or calcium ions and SiO_4 groups that give these minerals their hardness and low solubility in water. Chemistry handbooks commonly refer to these minerals as "insoluble," but if you have the patience of a geochemist, you can determine the almost imperceptible rates of dissolution in the laboratory, measuring them over weeks and months, or even years—and hereby joining the unofficial Society for Long-term Experiments. [I once monitored the dissolution of quartz in distilled water at room temperature for two years and watched the silica in solution climb to only 1 ppm (part per million) in all that time.] We could speed up the experiments by raising the temperature, but then we couldn't be sure that the

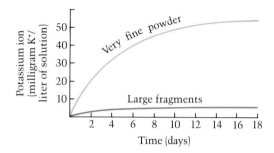

The rate at which potassium feldspar partially dissolves in water. The amount of feldspar that dissolves is measured by the amount of potassium ion liberated to the solution, which is measured as the number of milligrams of potassium per liter of solution. The feldspar dissolves at a fast rate in the first time intervals and then slows down until finally no further change can be detected. Grinding the feldspar to a fine powder makes it dissolve faster and allows more of it to dissolve.

same chemical dynamics were operating. The relative rates of dissolution of the three feldspars are determined by the relative proportions of sodium, potassium, and calcium: K-feldspar dissolves most slowly, Na somewhat less slowly, and Ca the fastest.

K-feldspar serves as a model for all the feldspars. Pure natural samples of this variety are readily available and have been extensively investigated. Dissolution experiments in the laboratory are guided by the observations of geologists who study the weathering of feldspar in the field. They had long ago identified the common reaction product of the weathering of feldspar in surface natural waters—rain, soil, and river waters—as the clay mineral kaolinite $[Al_2Si_2O_5(OH)_4]$. Because surface waters as well as ordinary solutions in the laboratory contain CO_2 in equilibrium with the atmosphere, the water is a dilute solution of a weak acid, carbonic acid, formed by the reaction $CO_2 + H_2O = H_2CO_3$. The dissolution equation is written with partially ionized H_2CO_3 as a reactant:

$$2KAlSi_3O_8 + 2H^+ + 2HCO_3^- + 9H_2O$$

$$= Al_2Si_2O_5(OH)_4 + 4Si(OH)_4 + 2K^+ + 2HCO_3^-$$

The solubility constant for this reaction,

$$K = \frac{[Si(OH)_4]^4[K^+]^2}{[H^+]^2}$$

at 25°C is extremely low, about 10^{-6}. At equilibrium (which takes months at room temperature to reach) in neutral (pH 7) water, the amount of K^+ in solution would be about 10^{-2} mol/L (moles per liter) and dissolved silica about 10^{-4} mol/L. As the feldspar reacts, it is etched in patterns controlled by its crystal structure and defects. The alteration product, kaolinite, builds up along the corroded paths of solution. Remarkably, the patterns of dissolution of the solid are much the same in the laboratory as in the natural soils, which gives us some confidence in the artificially simple experiments' application to natural weathering.

In these experiments, the reaction and its kinetics tell us which chemical factors determine the yield of the feldspar alteration reaction. The more acidic the solution, the greater is the extent and speed of reaction. With increase in temperature, the reaction

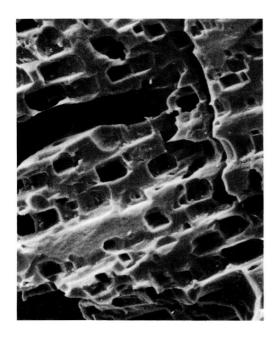

Scanning electron micrograph of a weathered feldspar produced by dissolution in soil water. The rhomboidal shapes of the etched holes reflect the crystal structure of the feldspar.

speeds up and the equilibrium constant increases. The more water, the more the reaction is displaced toward dissolution. If we remove the dissolved products of the reaction from the reaction vessel, say, by dialysis, we continue to force the reaction. With these experimental procedures and results in mind, we turn to the natural situation.

CHEMICAL WEATHERING IN NATURE

In the field, we can think of the feldspar-alteration reaction as proceeding irreversibly with a constant partial pressure of CO_2 in the atmosphere [normally about $10^{-3.5}$ atm (atmospheres)] and the reaction products, $Si(OH)_4$ and K^+, being drained off as rain water runs off an exposed surface of granite. An approximation of the rate of this reaction is given by the buildup of a chalky, whitish film of kaolinite on the feldspars of granite gravestones. A thickness of a few micrometers may be built up in 100 years, depending on rainfall and temperature (the rates of laboratory dissolution are consistent with this amount). Given this correspondence, we move on to the effects of water, acidity, and temperature.

The rainier the climate, the more water flows around the feldspars in a rock, and the faster the dissolved K^+ and $Si(OH)_4$ are carried away. The humid climate of a rain forest weathers feldspar far faster than the arid regime of a desert. The acidity in weathering environments is increased by organic acids produced by a metabolism of plants, algae, and bacteria; higher CO_2 pressures can also increase the carbonic-acid content. Some years ago, when geologists and soil scientists began to measure the pH and dissolved CO_2 of soil waters, they discovered that respiration in the soil by plant roots and by teeming underground worms, insects, and microorganisms can raise the CO_2 level by an order of magnitude. The biological connection is nowhere more apparent than in its important effect in weathering.

The temperature effect in weathering would seem to be simple. Because most materials are more soluble and dissolve more quickly as the temperature rises, we look for the contrasts between the highest and the lowest temperatures. The extremes of temperature at the Earth's surface are astonishing. The hottest temperature reliably reported is 54°C in Death Valley, the coldest,

−87°C in the USSR's Vostok experiment station in Antarctica. But that extreme temperature difference of 141°C exists only for short times between a few places. The mean annual temperature is a better measure for the extremely slow reaction of feldspar dissolution. The difference between the mean annual temperatures of the tropics, about 25°C, and those of high latitudes, about −20°C, is not so great. The 45°C difference accounts for relatively small changes in both equilibrium and reaction rate, and it is certainly not commensurate with the great differences in feldspar weathering we see between tropics and poles.

We can resolve this apparent contradiction by correlating both weathering and temperature with a third variable, the activity of living things. Small changes in mean annual temperature bring large differences in both vegetation and the animal population that feeds on it. Weathering processes are affected most directly by the change in vegetation from polar tundra (lichens, mosses, and grasses) through high-latitude temperate taiga (pine, fir, and spruce) and warm temperate deciduous forests to low-latitude tropical rain forests. Humidity increases with temperature in this progression, though not everywhere; for example, the extremely humid climate of the rain forest on the Pacific Coast of Washington's Olympic Peninsula is cool, an unusual contrast to the more typical hot, humid tropical rain forests. Again, the biosphere plays a dominant role in speeding up weathering. Differences in humidity clearly create differences in vegetation that we see magnified in the extremes between desert and rain forest, for even though most deserts are hot, the lack of water for much of the year prevents anything like the abundant vegetation of the rain forest. Thus, weathering is slower in a hot, dry desert than it is in a cool, rainy forest.

The weathering of other silicate minerals, such as pyroxenes ($FeSiO_3$ is one variety), takes place through the oxidation of reduced metals as well as through dissolution. The rusting automobile gives an appropriate analogy. As the iron oxidizes and hydrates, rust—a mixture of hydrated iron oxides—flakes off and leaves corrosion pits and holes. As iron silicate minerals weather, the iron oxidizes, forming the characteristic brownish rust on weathered surfaces. As the iron leaves the silicate, the crystal structure falls apart. When that happens, the remaining silica dissolves in rain water and is flushed away. The only remnant of the

Black volcanic-rock sands formed by fragmentation and weathering of basaltic lavas in Hawaii. *Left:* Sand grains are fragments of basalt and grains of olivine and pyroxene. *Right:* A black sand beach at Polulu, Hawaii.

iron silicate mineral may be a stain of rust on the rock face. A piece of steel buried in soil, if left long enough, will vanish. The soil, itself a product of weathering, speeds up the weathering process of the rock it covers.

The rates of dissolution of silicates vary widely, with the ones containing reduced metals such as iron among the fastest. The mineral olivine $[(Fe, Mg) 2SiO_4]$ dissolves so rapidly that it rarely survives as a sand grain in temperate climatic regions—we find large amounts of olivine sand grains only on some black sand beaches of volcanic islands such as Hawaii, where the supply of olivine from basaltic volcanic rocks is great. In contrast, quartz, one of the most abundant minerals of all kinds of rock, dissolves so slowly that many more years than the two I was willing to spend in my own long-term experiment are required to dissolve 1 mg (milligram) in a liter of water at room temperature.

These large differences in dissolution rates reflect strong differences in chemical stabilities of the silicate minerals. These differences cause wide variations of survival rates of the minerals of a weathering granite, with quartz surviving the longest and olivine the shortest. This gives us a workable index of weathering: the ratio of any mineral to quartz in the weathered residue compared with the ratio in the parent rock. A freshly blasted surface of a road cut in the warm, humid climate of Georgia might show the parent granite to have a ratio of amphibole (a fairly unstable Ca, Na, Mg, Fe, Al silicate) to quartz of about 0.1. After some hundreds of years the amphibole/quartz ratio would be significantly smaller, perhaps around 0.05. After thousands of years, most of the amphibole might be gone, but the feldspar ratio, initially about 0.5, might be reduced to 0.3. We now have the beginnings of a quantitative measure of the intensity of weathering in a parameter directly related to climate and topography that we can use if the weathered residue—sand—is preserved geologically. But the story does not end here. How are the constituent crystals of granite that are resistant to decay broken out of the rock to form sand grains?

FRAGMENTATION TO SAND GRAINS

Sand grains are the crystals of a granite in another guise. They are the products of weathering, fragmentation, and dissagregation, which combine to make a particulate sediment out of eroding rock. The first step in this process is fracturing of the rock, which provides access to the water that will corrode the unstable minerals. All rocks that were once buried deep in the Earth's crust are cracked to some extent from the stresses imposed during formation and deformation. Cracks exposed at the surface tend to open up into fractures only of the order of micrometers, or even less, but infiltrating waters quickly widen them into the obvious wide breaks of outcrops. At the outset these fractures divide the rock into the large blocks that accumulate at the bases of cliffs or into smaller pieces that will form gravels.

Most sand grains are made in another way, by the rock disaggregating along crystal interfaces. The stresses imposed on rocks buried at great depths in the Earth strain the rock along intercrystalline boundaries as well as over the larger-scale pattern of fractures. A boundary between a feldspar crystal and its neighbors is vulnerable to microcracking and invasion by water. It is doubly vulnerable because as the feldspar starts to change into kaolinite

Weathered outcrop showing the enlargement of joints and fractures by chemical and mechanical weathering and the production of abundant mineral and rock fragments mantling the slopes by the disaggregation of the rock.

along the crack, it weakens the strength of the interfacial bonds between crystals. The rock starts to fall apart into its constituent crystals when enough of these cracks have enlarged. Processes such as the freezing of water or the pressure of plant roots growing in the opening crack can physically wedge the grains apart. The result of these concerted processes is a rubble of disaggregated crystals, some of which are partially altered by chemical weathering and mixed with larger fragments and blocks. It is no coincidence that in repeated comparisons, we find a strong correspondence between the size distributions of sand grains and the crystal size distributions of parent rocks.

Not all rocks are composed of a set of mineral crystals of sand size: volcanic lavas are largely silicate glass, and many other rocks are made up of crystals so fine that they would make parti-

Basalt cliffs and seastacks and the beach sand partly derived from their erosion. Pistol River State Park, Oregon.

cles of 0.05 to 0.001 mm in diameter, material of clay and silt size. Observation and experiment have shown that until chemical weathering is far advanced, the cohesive forces between very fine crystals and within the glassy matrix are too strong to allow the rock to be broken by splitting along intercrystalline boundaries. Sand grains that are produced from these rocks by small-scale fracturing and abrasion are polycrystalline, small representatives of the large rock mass.

A walk along some beaches of the Oregon coast offers us not only the spectacular delights of the imposing basalt cliffs and strange rock pillars rising above the waves but also a fine lesson in the production of dark sand grains. These basalt cliffs were extruded as sea-floor lavas not many millions of years ago—at an early stage of the evolution of the Cascade Mountains by the subduction of the Juan de Fuca plate under the North American plate (see Chapter 7). The basalts had congealed rapidly, quenched under cold seawater, and accumulated strains within the glassy matrix. Later uplift of the Earth's crust produced the coast ranges, and erosion then started to work on the basalt. During weather-

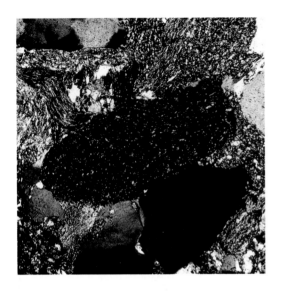

Photomicrograph of a sandstone rich in rock fragments. Most of the grains that have a speckled appearance are finely crystalline fragments of shales, schists, and slates that show well-rounded outlines. Volcanic rock fragments are common (see dark elongate grain at center).

ing, strain patterns controlled the fracturing on large and small scales, producing both large blocks and sand grains.

Although single-mineral sand grains, such as quartz, are the raw material sedimentologists work with most often, multi-mineralic rock fragments are special treasures for the source-rock detective. They give the rock type directly, sparing us the task of mentally reconstructing the rock from single-mineral sand grains. The pebbles of a gravel are even better, almost as good as an outcrop.

TECTONIC CONTROL OF SAND PRODUCTION

Whether sand grains are single crystals or rock fragments, the processes of fragmentation that make them are in tenuous balance with chemical decay. Where decay is fast and fragmentation slow, unstable grains such as feldspar and pyroxene are lost before they are split out as sand grains. Where fragmentation is fast compared with decay, all the different kinds of minerals contribute to the sand-grain population. In this population, the ratio of fragmentation, or mechanical erosion, to decay, or chemical weathering, is something we can measure. At the same time, we can relate that ratio to the geologic controls on weathering and erosion.

The measurement of the ratio is simple. Why not take the ratio of unstable minerals to quartz, our index mineral for slow weathering? A high ratio of feldspar to quartz indicates mechanical erosion, with little time for chemical weathering to operate. A pure-quartz sand, on the other hand, indicates deep chemical weathering and slow mechanical erosion. A trained geologist can quickly determine the quartz/feldspar ratio by counting several hundred sand grains in a thin section of rock under a polarizing microscope.

Now we come to geological interpretation. Mechanical erosion is subordinate to chemical weathering in all places where—for reasons of temperature, topography, humidity, and vegetation—the rates of decay, although they may be slow, are still faster than the rates of fragmentation, the cutting of river channels, the erosion of steep cliffs, or the general slow lowering of the land surface by rain and wind. We can observe this kind of regime favoring chemical decay in warm, humid lowlands. To see where mechan-

The comparative effects of climate and topography on the relative amounts of mechanical erosion and chemical weathering. Chemical and physical breakdown of the source rock is reflected by the quartz to feldspar ratio. Chemical dissolution is most intense in warm, humid lowlands with abundant vegetation, least in cold, arid, mountainous regions.

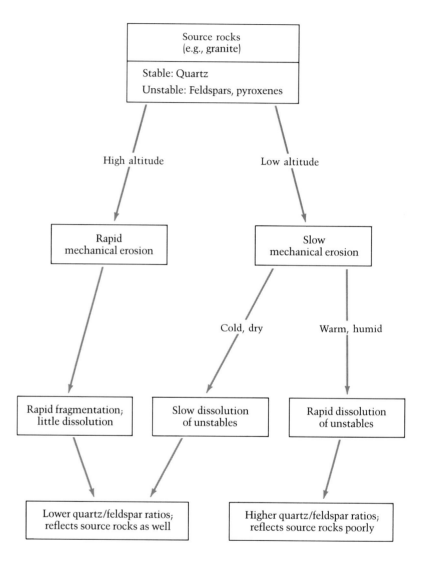

ical processes dominate, climb up to the top of the nearest mountain, especially those in polar regions. There the cold, dry, unvegetated slopes erode more rapidly than they decay. But what would you expect from a mountaintop in the tropics? An analysis of river waters draining the tropical Andes Mountains indicates extensive chemical decay in the abundant products of weathering—the dissolved sodium and other ions and silica from dissolution reactions. Despite this extensive decay, the physical erosive processes on the high mountains still overwhelm the chemical ones, and the sand grains coming down the high mountain streams reflect fairly well the original composition of the eroding rocks (see the figure on page 30).

Once we establish topography as the important control on erosion and weathering, we immediately have a guide to larger-scale Earth processes. Mountains, the first-order topographic features of the continents, are the products of tectonic activity. The motions of oceanic and continental lithospheric plates may lead to continental collisions, such as the one between India and Asia, which formed the Himalayas. Or they may lead to oceanic–continental convergences, such as that of South America and the Nazca plate of the adjacent Pacific, which formed the Andes. Ultimately, the different motions of plate tectonics, which produced the great mountain ranges of the past, can be inferred from sandstone compositions. A low quartz/feldspar ratio automatically suggests high relief that is produced by mountains thrown up by tectonic forces. The summer vacationer in the low hills of the Taconics of eastern New York State could read their origin from the mineralogy of the sandstone bedrock beneath the soil. They are the long-worn-down remnants of a high mountain range that was thrown up 450 million years ago. At the same time, the preservation of unstable minerals in feldspar-rich sandstones provides precise information on the rock types exposed in the mountains. As we will see in Chapter 7, we can use the rock types to infer just what kind of plate tectonic motions produced those mountains.

In the opposite case, where quartz/feldspar ratios are very high— because almost all the unstable minerals have been removed by chemical weathering—we read information of a different kind. With these high ratios, we conclude that intense weathering, presumably of humid lowlands, indicates tectonic quiescence. The source area was far removed in space or time from plate-tectonic collisions. We cannot tell much about the original source-rock composition if the sand derived from it is all quartz—there is a limit to how closely quartz varieties correspond to rock compositions. And we cannot be sure that these quartz grains were not derived by erosion of an older quartz sandstone exposed at the surface. In fact, we cannot be absolutely sure that a preexisting quartz sandstone was not the only rock type exposed in a mountain belt. That possibility is slim, but given the diversity of rock types normally exposed in a mountain chain, it occasionally happens. We can resolve this dilemma by taking a different tack, setting aside questions of mineralogy and concentrating instead on grain textures.

THE SHAPES OF SAND GRAINS

Sand grains formed at the sites of weathering are atypical of most sands we are familiar with. Grains at weathering sites, liberated from a mass of granite, are angular, exhibiting the sharp edges and corners of the crystal shapes or fractured particles of the parent rock. Most of the sand grains we see on the face of the Earth are more rounded, their edges and corners smoothed by abrasion as running water or wind moves the sand from its birthplace and the grains knock against each other like billiard balls. About 50 years ago sedimentologists learned how to quantify roundness or its inverse, angularity, by taking the ratio of the average radius of curvature of corners to the radius of the grain: a grain of radius 0.5 mm with many small, sharp corners with an average radius of 0.05 mm would have a ratio of 0.05/0.50, or 0.1; on a smooth sphere the single "corner" is identical with the grain and the radii are the same, giving a ratio of 1.0. Originally the ratio was determined by laborious measurements of individual grains. With the availability of computers, roundness can be determined efficiently by fitting a Fourier series to the periodic function of the radius of the grain as a function of angular distance about the center of the grain. However determined, the roundness is a function of abrasion induced by transport. As it turns out, roundness of quartz—the standard mineral used in roundness studies—increases extremely slowly with distance.

Experiments with rotating devices that simulate the rolling about of sand grains on the floor of a stream or under ocean waves, have shown that thousands of miles of transport are required to achieve even moderate rounding. Thus the angularity of the sands of Surinam rivers reflects their short transport distance. On the other hand, a beach, where sand moves in and out with each wave, is the ideal place for rounding sand grains if they stay there for any length of time. Now the picture becomes more com-

Angular and rounded sand grains. *Left:* Angular, olivine-rich sand, South Point, Hawaii. *Right:* Rounded quartz sand, Florida.

plicated: to interpret the roundness in terms of transport distance, we must also know the transport agent. But even beaches hundreds of kilometers long seem, by themselves, unable to produce extremely well-rounded grains from angular quartz crystals freshly weathered from a granite—and certainly not the almost perfectly smooth, spherical grains we can see under the microscope when we look at sands such as the St. Peter Sandstone of northern Illinois and southern Wisconsin. The source of extremely well-rounded sands turns out to be preexisting sandstones that were eroded. Quartz sand is so stable that it may be recycled through erosion, sedimentation, burial, and uplift many times and still retain its identity as a grain. Through these cycles it is repeatedly abraded during transport; the final result is a rounded grain that stands as a record of all the cycles of mountain building and erosion that it has gone through.

Each cycle of erosion and deposition also takes its toll in the disappearance of chemically unstable minerals. If a certain amount of feldspar is weathered to kaolinite during the erosion of

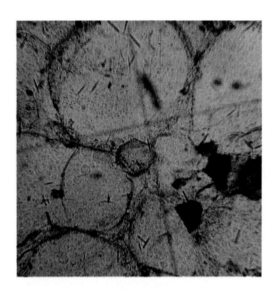

Photomicrograph in plane-polarized light of rounded detrital grains of quartz. These grains were rounded by abrasion during several cycles of erosion, transportation and deposition. The grains are cemented by the later addition of post-depositional quartz between the grains.

a granite, then even more will disappear when the sand derived from that granite is itself eroded. After several cycles little may be left other than quartz.

We face a choice in interpreting the origin of a pure quartz sand: did it come from the deep, intense weathering of a granite or from successive erosion deposition cycles of sandstones derived secondarily from the original granitic source? The roundness of the grains is the clue that leads us to the multicycle explanation.

One might think that the operation of the geological machine would make either possibility equally probable. After all, there are great quantities of both granite and sandstone exposed to erosion at any time. Furthermore, the longer the geological time available, the higher the probability that sandstone will be cycled repeatedly. We should expect, then, that geologically much younger sandstones would tend to be more quartz rich: but just the opposite is true. This paradox, gradually becoming known in the last half-century, is starting to be answered as we take a new look at the evolution of this planet, a subject we return to in Chapter 10.

RECONSTRUCTING SOURCE-AREA GEOLOGY

In any case in which there is little modification of the debris of erosion by weathering and transport, we are able to say a great deal about the rocks of the region being eroded to supply the detritus. Simply count the minerals and rock fragments making up a sand and compare them with the kinds of minerals and their relative proportions in various parent rocks. For example, finding a sandstone consisting of 55 percent feldspar, 25 percent quartz, 10 percent biotite (a dark mica, $[K(Mg,Fe)3AlSiO_3O_{10}(OH)_2]$), and 10 percent amphibole immediately suggests a granite. However, demonstrating that the sandstone is evidence of eroding granitic mountains is not the end of the story. Plate-tectonic theory predicts where such granitic mountains form and what relation they have to plate motions and geologic history. Such terranes, regions of characteristic rock formations, might be found along the borders of continental rift valleys or along collision zones between two continental plates. Discriminating among plate-tectonic settings requires that we take into account the relationships among several variables: tectonics, the formation of igneous and meta-morphic rocks, patterns of sedimentation, and the specific geological history of the region. We discuss this subject in further detail in Chapter 7.

Nature is rarely so simple as to provide us with the ideal un-weathered sediment to study. Instead, we usually see evidence of intermediate amounts of weathering, and we then have to evaluate the differential effects of climate and topography. Another complexity is the mixing of material from several source terranes. Source terranes are rarely homogeneous granitic areas. Metamorphic rocks are common in association with igneous rocks. Sedimentary rocks may be exposed along the borders of the source terrane. Rivers draining some of the Appalachian Mountains carry detritus from eroding sedimentary rocks of the Valley and Ridge region mixed with igneous and metamorphic debris from the lower hills of the Piedmont to the east. The climate in the northern parts of the Appalachians is temperate and weathering is moderate, but warmer, more humid climates to the south, in the Carolinas and Georgia, result in deeper weathering. Northern and southern rivers draining much the same kind of terrane carry different loads because of the climatic differences.

In ancient rocks we have to infer both climate and topography. How do we choose? One way is to use paleomagnetism to tell us the paleolatitude, or original latitude, at the geologic time of concern and work from that to suggest the range of climates. In many reddish brown sandstones, we find a remanent, or residual, magnetism that differs from that expected of the Earth's present magnetic field. That magnetism was imposed by the Earth's magnetic field at the time of sandstone deposition. The magnetism resides in magnetic iron oxides chemically deposited as coatings on the grains soon after deposition—so soon geologically that they are essentially contemporaneous. The remanent magnetism tells us where the magnetic poles were at the time of deposition, thus giving the ancient latitude of this region of a plate. Armed with a paleolatitude, the geologist can make a first estimate of a likely climate. Latitude coupled with a paleogeographic map showing the positions of shorelines and mountain ranges leads to a more refined assessment. Putting mineralogy, texture, and paleomagnetism together to study paleoclimates is now an active field of research in the geology of sandstone.

CHAPTER THREE

Sand travels

Sand grains are travelers. All it takes is a moderate wind to blow the small ones, and almost any respectable water current can carry medium and coarse grains. Millennia ago our ancestors were aware that rivers carried sand downstream, that desert winds blew sand into dunes, and that the waves washed sand back and forth along the strand. We now know that the ubiquity of air and water currents on the face of the Earth is responsible for the widespread distribution of sand over the continents. We can in fact infer the existence of ocean currents, even those too transitory to measure directly, from the distribution of sand on the sea bottom.

Almost as soon as the earliest natural historians discovered how sand was transported, they must have related this phenomenon to current velocity. The faster a stream, the larger are the grains that it can carry and the greater is the volume of sand it can transport. From these observations our first deduction became possible: the coarse-grained sands of large sandbars on a river must have been deposited when the current was very fast, perhaps when the river was in flood. That stage of deduction was where we stayed until the science of fluid motions, fluid mechanics, came into being in the seventeenth century. Today, sedimentologists experiment with laboratory flumes and correlate their results with observations of natural situations. They build models of sand-transport mechanics, the ways in which fluids transport solid particles, based on modern knowledge of fluid flow and the interactions of the fluid with particles in transport. From such studies have flowed interpretations of the fluid-flow conditions, or the relationship of forces and motions in fluids, under which ancient currents operated, in terms not only of velocity but of density, depth of water, and other parameters, both fluid mechanical and geological. Armed with this knowledge, a sedimen-

Sand dunes advancing from the beach drown trees in Oregon.

Sandbars on a straight section of a river. The bars form in response to variations in current velocity that are similar to those in a meandering river.

tologist in the field can reconstruct the current that transported the sand, whether it is a tropical meandering stream, ocean waves on a beach with a high tidal range, or a submarine current flowing down the continental slope under thousands of meters of water.

To get to this point, we start with a microscopic view of sediment-transport mechanics: how a fluid flow picks up and carries a sand particle along with it. The laws of these mechanics alone allow us to predict when a sand particle of a given size, shape, and density will be transported and when it will be dropped. Moving from the theoretical to the natural occurrence, we can explain gold nugget deposits in streams—and how panning for gold works. But this individual-grain approach takes us only so far. To understand how cross-bedding and ripplemarks form, we need to increase the scale of our approach by studying aggregates of millions to billions of grains. (The 8-m flume in my laboratory contains, for most experiments, about 15 million grains, each of which is 1 mm in diameter.) This is the scale of an individual small outcrop of sandstone, which is comparable in size to a trench of a few meters cut with a spade along the length of a river sandbar. At this scale we can start to understand some fundamental dynamics of transport: for example, the waxing and waning of a river in flood and the way the highs and lows produce sequences of sand beds with different sedimentary structures. Here is the major clue to deciphering the sedimentary environment in which the sand was deposited.

In larger-scale natural environments, where we observe domains of thousands of cubic meters containing many trillions of grains along a stretch of beach, we begin to see the effects of the variability of currents over larger areas. Changes in currents make sands differ from place to place along a road cut or between outcrops. Finally, at an even larger scale, we can map entire river drainage systems and see how the current changes as the river descends from its mountain source to the lowlands near its mouth. On this scale we can map paleocurrents, the larger patterns of sediment transport that reflect the major architecture of continents and ocean basins. No matter what the scale, all our knowledge rests on understanding how the physics of fluid flow influences the transport and deposition of particles in that flow.

At each level in this hierarchy we need to use the appropriate scale of experiment and observation. We can experiment in the laboratory only on scales from centimeters to a few tens of meters. The brave new world of deliberate scientific experimentation with artificial rivers many hundreds or even thousands of meters long awaits the time when scientists, engineers, and government-funding agencies have the willingness to experiment with very large systems. In the meantime, for these and even larger scales we rely on perspicacious observation.

FLUID FLOW AND SEDIMENT-TRANSPORT MECHANICS

Fluids are changeable. They change their shape to fit the container. Whether they are liquids, which are relatively incompressible, or gases, which are highly compressible, fluids can be pushed through orifices or sucked into tubes. Their molecules are free to wander about each other, so that any orderly structure they take is transitory. If we impose a stress on a fluid, it will instantly relax when we remove the stress.

Yet fluids show constancies too. All fluids resist deformation, a property that is vital to our understanding of their behavior, especially with respect to sediment transport. In general, fluids exhibit constant density and other physical properties even while they are deforming and flowing. (When we come to the largest

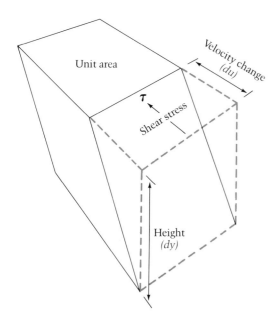

Unit area

Velocity change *(du)*

τ

Shear stress

Height *(dy)*

A simple geometrical illustration of the shear stress between two layers of a fluid. As the top layer flows over the bottom layer, the unit volume is deformed from a cube into a rhombus. The shear stress is equivalent to a force that produces a change in velocity, *du*, over height, *dy*.

scale of fluids on Earth, the oceans and atmospheres, we will see that because of pressure and composition changes, even these fluids are not constant.) In considering fluids in relation to sand transport, we will restrict ourselves to appropriate scales: to liquids and gases in small laboratory tubes, to rivers, to small domains of the ocean such as shallow waters near a shore, and to the volumes of air within tens of meters from the ground. Fluids at all these scales show constant physical properties, and all are large enough to allow us to ignore the molecular scale and treat the fluid as a continuum, ascribing properties to "particles" of fluid at a point.

Returning to the idea of deformation, we can use it to define a property distinctive of fluids, a characteristic of a particular kind of fluid, such as water, melted butter, or tar. That property is the viscosity, or resistance to deformation. We commonly measure it by seeing how long it takes a given volume to flow through a small orifice, but you can visualize it more simply as the shearing deformation of a unit cube into a rhombus. The shearing deformation is caused by a force, the shear stress, acting tangentially on a volume of fluid. As Newton showed in his *Principia* in 1687, the tangential shear stress of a movable plate moving over a layer of fluid based on a fixed plate, the upper and lower surfaces of the cube in the figure on this page, generates a velocity gradient. The gradient, *du/dy*, is proportional to the shear stress, and the proportionality constant is μ, the dynamic viscosity.

$$\text{Shear stress} = \mu \frac{du}{dy}$$

The shear stress of a fluid acting over the bottom of a river bed is responsible for entraining, or picking up, sand grains from the bed and carrying them along in the flow. Because shear stress is directly proportional to velocity, it is apparent why the ancients could already relate sediment transport to velocity. Although the viscosity is a constant everywhere in a fluid of a given kind, it is dependent on intermolecular forces and thus on temperature. The viscosity of water changes significantly as the temperature falls, from 1.002 at 20°C to 1.787 cP (centipoise) at the freezing point; 1 cP equals 0.01 P (poise), which is defined as 1 dyne second per

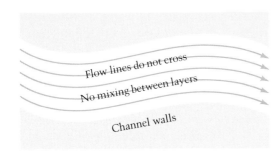

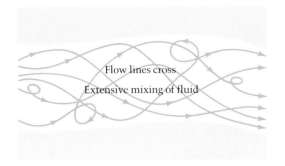

Top: Laminar flow of a fluid between two solid channel walls. In this type of flow, flow lines do not cross, and there is no mixing of fluid between layers. *Bottom:* Turbulent flow of a fluid between two solid channel walls. Flow lines in this type of flow cross in a complex pattern; unlike laminar flow, no definite layers of flow can be discerned, and extensive mixing of the entire fluid mass occurs.

centimeter squared. The increase in viscosity gives us a reason for the increased sediment-transport capability of a stream of melt-water in the early spring over that of the same stream in midsummer. In the fluids sedimentologists worry about, the shear stress increases linearly with velocity gradient, a characteristic of all newtonian fluids, named for the discoverer of the fundamental law of viscosity.

As a fluid deforms, curving around bends in a narrow plastic tube, it may follow an orderly pattern in which all flow lines are parallel and no mixing occurs between layers. These laminar flows are characteristic of slow, viscous fluids. But the river flowing in a broad arc around a meander bend is more likely to display turbulent flow, in which the layers of water mix in eddies, swirls, vortices, and boils, constantly transfering both material and momentum from place to place in the flow. Turbulent flows resist deformation more than laminar flows, so that turbulent viscosity appears larger, but not uniformly. In contrast to the true or dynamic viscosity, the viscosity of a turbulent flow is dependent not only on the kind of fluid but on the degree and kind of turbulence. Thus, a highly turbulent flow of a low-viscosity fluid might be more resistant to deformation than a more tranquil flow of a higher-viscosity fluid. This leads to the formulation of a new kind of viscosity, the eddy viscosity, that we add to the dynamic viscosity to give the total relation between shear stress and velocity. Where the dynamic viscosity, because it is a function of a fluid's intermolecular forces, is a constant for a given fluid, the eddy viscosity is a variable that measures the turbulence.

What makes the difference between laminar and turbulent flows? In 1883, Sir Osborne Reynolds, an English physicist, reported a classical series of experiments that showed how the transition from laminar to turbulent flow occurred as the velocity increased, the viscosity decreased, the roughness of the boundaries of the flow increased, and the flow became less narrowly confined. He combined these factors into a dimensionless parameter relating velocity, geometry of flow (defined by pipe diameter by engineers and depth of a stream by river hydrologists), dynamic viscosity, and density:

$$\mathbf{R} = \frac{\text{velocity} \times \text{length}}{\mu/\text{density}}$$

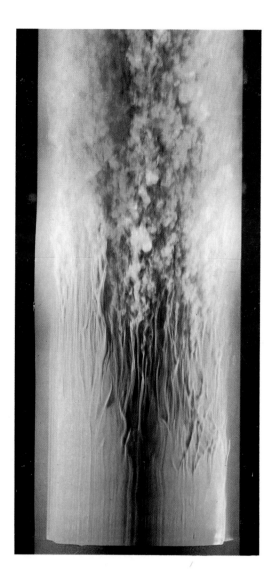

The transition from laminar to turbulent flow in water along a flat plate, revealed by dye injection.

The critical Reynolds number (**R**) gives the value at which the flow changes from laminar to turbulent or vice versa. The Reynolds number can be shown to be the ratio of inertial forces of the flow to the viscous forces of the resisting fluid. Where viscous forces dominate and the Reynolds number is low, the flow is laminar. In turbulent flows, inertial forces swamp the resisting viscous forces. In natural watercourses with relatively rough boundaries, the critical Reynolds number is about 500, whereas in smooth pipes it is about 2000. Most streams, the surface layers of the ocean stirred by the winds, and the winds close to the ground are turbulent. We can see laminar flows in small rills with thin flows of a few millimeters but rarely elsewhere on the natural surface of this planet. The prevalence of turbulent flows accounts for the sediment-transporting capability of the many different sorts of natural currents found in the environments of the Earth's surface.

Thinking about transport of sand grains by current inevitably points us to the interactions between a current and the sandy bottom over which it flows; here, along the bottom, sand grains are picked up and dropped to rest when the current slows. The bottom, the boundary of the flow, both constrains the flow geometrically and, by acting as a retardant, slows it. Early in this century, Ludwig Prandtl devised a new way of working with the effects of a boundary on a flow: simply separate the main flow (the part unaffected by the boundary) from the boundary layer (the zone where the fluid is retarded by the frictional resistance of the boundary). Whereas the main flow is invariant with distance from the boundary, the boundary layer shows a strong decrease in velocity as the boundary is approached. In this zone the flow picks up loose grains with a velocity dependent on its position in the boundary layer; this is always less than the velocity of the main flow. The roughness of the boundary strongly influences the boundary layer. A smooth bed of fine sand is not much of a retardant, but as the bed becomes rippled or covered with small dunes, it offers more resistance by deforming the boundary. The same is true of the coarse gravel or boulders of a mountain stream. These effects of the boundary are so strong in natural watercourses that most streams and rivers are entirely in boundary flow: the entire flow is affected by the boundary and there is no main flow.

The drag, or resistance of the boundary to the flow, plays an important role in sediment transport. Here we can conveniently

The frictional resistance of the boundary of a flow produces a boundary layer with retarded flow adjacent to the boundary. At some point above the boundary, the retardation becomes negligible, above which the main flow has a constant velocity at all heights. Depending on the velocity and the smoothness of the boundary, the boundary layer may be very thin or may comprise most or all of the flow.

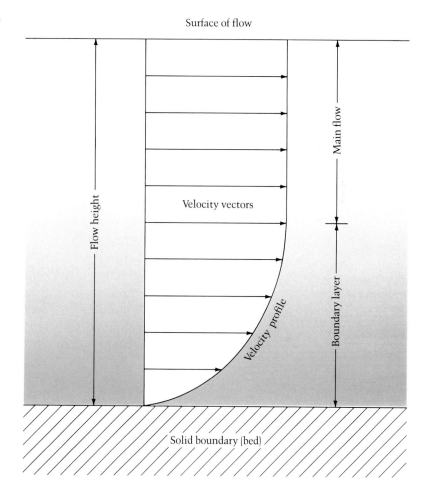

think of particle transport as a balance between entrainment, the picking up of a particle from the bed, and settling, the dropping of a particle to the bed. A particle stays suspended in the flow and is therefore carried along with it if the rate at which it settles—the settling velocity—is greater than the rate at which turbulent-flow forces keep the particle up in the flow.

Analyzing settling velocity has been straightforward since G. G. Stokes showed us how to do it in 1851. We can think of settling velocity in terms of the gravity force of the particle falling, lessened both by the bouyant force of the fluid on the particle and by the drag of the particle, the surface of which acts exactly as any boundary would, for example, a pipe or a stream bed. Given the resistance of the fluid to the fall of the particle, at some point the gravitational acceleration of the falling particle will be countered by the drag. From that point on the grain will fall at its terminal velocity (zero acceleration).

Gravitational force (weight)
= bouyancy force + drag force

We can easily determine the weight of the particle and, from the densities of fluid and particle, the bouyancy force. By experiment and empirical analysis, we can express the drag force in terms of the size of the grain, the density of the fluid, and the settling velocity. The latter comes into play because drag is a function of the velocity of either the flow past a boundary or, as in this case, the flow of a particle past a fluid. We also need a "fudge" factor, the drag coefficient, which takes into account the shape of the grain and its surface roughness. For example, for spherical, smooth grains, an expression for the drag force, F_D, is

$$F_D = C_D\left[\left(\pi\frac{d^2}{4}\right)\left(\rho_f\frac{u^2}{2}\right)\right]$$

where C_D is the drag coefficient, d is the diameter of the grain (the most convenient measure of size), ρ_f is the fluid density, and u is the velocity. From these equations we can calculate the settling velocity of a particle, which comes out to be

$$u = \frac{1}{18}\frac{d^2g}{\mu}\left(\rho_s - \rho_f\right)$$

where μ, as before, is viscosity, g is the value of gravity, and ρ_s is the density of the solid (grain).

In its simplest form, this equation for the typical case of a quartz sand grain falling through pure water at 20°C is

$$u = kd^2$$

where k is 8.9×10^3. The constant k lumps together all of the variables of viscosity, drag coefficient, density, and the value of gravity. We now have a measure of velocity, one of the important variables of the flow, in relation to the size of the grain carried; this measure allows us to predict in a general way the grain size of a deposit from a given velocity flow. Even more important to a geologist, we can reconstruct the velocity of the flow from the grain size. The relation between velocity and size is useful in still another way: the practical measure of size for a large aggregate of grains. A quantity of sand is allowed to settle through a water-filled cylinder, falling on a pressure transducer connected to a chart recorder. The automatically produced graph of the pressure increase with time is converted to an average size.

The settling-velocity equation can also explain the formation of placer deposits, the segregations of gold, garnet, or titanium oxides, in some places found in minable amounts along beaches and rivers. Grains with higher density, such as the minerals found in placers, have higher settling velocities than quartz grains of similar size. If all the grains are the same size, the high-density grains will settle out more rapidly to form a more or less pure fraction of that mineral. Placers of this kind are common along such beaches as those of the northern Atlantic coast of North America, where grains of garnet, a fairly dense silicate (density about 3.8 g/cm^3), are found in purplish ribbons along the head of the beach. The effects of particle shape are demonstrated by the micas, the thin, flat colorless or brownish sheets found in so many igneous and metamorphic rocks. Because the mica flakes settle slowly, compared with more spherical shapes, large flakes are found with fine sands or silts.

These parameters control the settling out of a stream. Now we can turn to the entrainment that gets particles up into the flow. The effects of turbulent flows are immediately relevant, for we know from observation that laminar flows entrain far less sediment than turbulent ones. Because flowlines of a turbulent flow cross each other and produce eddies, local regions of uplift occur. Where they impinge on the bottom, these regions will exert a force that pulls a loose grain up into the flow. Another effect is the torque (a rotation imposed by counter forces on opposite sides of an object) produced on a grain that slightly projects into the lower part of the boundary layer as the forward movement of the flow tends to move it forward, pivoting it about the adjacent grain.

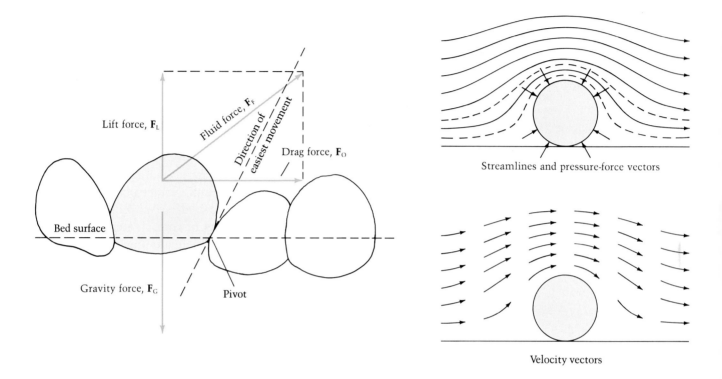

Streamlines and pressure-force vectors

Velocity vectors

Left: Forces acting on a sand grain on a sandy bed. The lift component of the fluid force on the grain ($\mathbf{F}_L$) results from the increase in velocity (and decrease in pressure) as flowlines converge over the grain. The horizontal drag force ($\mathbf{F}_D$) is the resistance to flow offered by the bed. These two forces combine to give the net fluid force, $\mathbf{F}_F$, that lifts or pivots the grain around a neighboring grain. *Right:* Flowlines and force vectors over a grain.

The effect of these forces is the entrainment of grains in the flow to varying heights above the bottom. There, it is immediately subject to the forces making it settle: gravity and the localized downward turbulent eddy. For most sand grains in the moderate velocity flows of typical streams, the net result is saltation, a jumping motion of the grain in which it follows a trajectory from the point of entrainment to the point of settling to the bed. Statistically, the grain may be at rest on the bed for a good part of the time. Yet at any instant, the stream may be carrying a significant amount of this saltated, therefore temporarily suspended, load of sand.

Anyone who has walked on a sand dune during a strong blow has experienced the saltation load directly. The jumping sand grains are everywhere, finding every crevice in your clothes and stinging your bare legs. The sand bed is covered with a cloudy layer through which you can see ripples only indistinctly. The

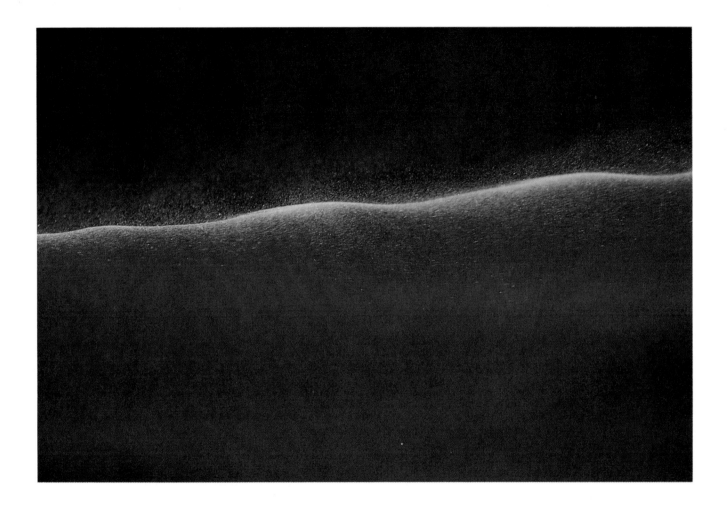

Saltating sand on a dune.

frequently waist-high top of the layer may be sharply defined. Your head is in the clear air while your feet are obscured. If you are unlucky enough to be caught driving in a sandstorm in the desert, your car's windshield may become as frosted as the groundglass screen of a camera. The frosted glass of soft-drink bottles in the sand is the product of the same sand blasting. Saltating sand can be an efficient erosive agent as well as a transporter.

Larger grains of sand, too heavy to be entrained, may move along the bed by rolling and sliding. This bed load becomes greater as the flow increases, reaching the point at extremely high flow rates where a good deal of the bed is in motion. It may be difficult at this point to distinguish where the bed ends and the suspended load begins. If you could keep your footing in such a flow over a fluidized bed, you would sink down as in a swamp. The technology of producing such fluidized beds has become important in transportation, where, for example, coal is crushed to fine sizes and mixed with a fluid to make a slurry that can be pumped at high pressures through a pipeline.

Unlike coarse grains and pebbles, the fine sand and silt particles at the opposite end of the sedimentary-particle size range are easily lifted up into the current. If small enough, these grains may stay permanently suspended in the flow, settling out only from the most tranquil flows. These grains travel down a river at the speed of the flow and are not held up by temporary rests on the bottom. Once suspended, the finest clay-size particles will not settle until the flow stops entirely, as it does when a river expands over its entire valley floor during a flood, then leaves still pools or lakes when it recedes and returns to its channel. We have all learned about the fertile farmlands of the Nile, nourished by the rich clay deposited during floods. But woe unto the farmer whose land is near a flood flow in which some of the sand normally carried only in the channel breaks out and gets distributed over part of the valley (see Chapter 4).

This brief visit to sediment-transport mechanics has introduced us to several attributes of sand grains that might be used to reconstruct the flow that deposited them. Obviously their size is important, and their shape too, as shown by the behavior of mica flakes. Surface roughness also enters the picture. For the bulk of sands made up of quartz, feldspar, and the many other common silicates that have about the same density as quartz, about 2.7 g/cm^3, we need not consider density. When we need to understand the deposition in placers of heavy minerals, such as garnet or the iron oxides, those with densities appreciably greater than quartz, we do need to take density into account.

SAND-GRAIN TEXTURES

To fully evaluate the effects of transportation, we need to look at grain textures. For most of this century, sedimentologists have

Size limits of sand grains

Size (mm)	Type
>4	Pebble, cobble, boulder
4–2	Granule
2–1	Very coarse sand
1–1/2	Coarse sand
1/2–1/4	Medium sand
1/4–1/8	Fine sand
1/8–1/16	Very fine sand
1/16–1/256	Silt
<1/256	Clay

labored at quantitating the distributions of sizes, shapes, and other textural properties of a sand, hoping for a quick fingerprint that would identify this deposit as a beach sand and that one as a river sand. The quick fix has not come yet, but we have learned a great deal in the course of the search. Most research efforts have been on size, an obvious first choice from the most elementary notions of fluid flow—and we had little more than that at the turn of the century. Reynolds' number and Stokes' law of settling velocity were newly invented and barely known in the physics community and Prandtl had yet to work on boundary-layer theory. Yet in 1898, Jon Udden, a professor of geology at Augustana, a small midwestern college, first published statistical analyses of sand-grain sizes in windblown deposits and later went on to compare them with river and beach sands. Udden had already anticipated the direction and substance of much later work.

How does one define "size" when sands are a mixture of sizes and each grain has its own shape, many of them far from the ideal sphere of fluid mechanics? The way fluid flow affects sand grains really depends on how heavy the particles are, so weight is the best thing to measure. Measuring the individual weights of thousands of sand grains to get a good average would have been tedious, so Udden decided, following the example of soil scientists, to use a series of sieves. In doing so he invented the forerunner of the modern size classification system, a geometric grade scale based on the square root of 2, as shown in the table.

This geometric scale has the advantage that $\frac{1}{4}$ mm made very little practical difference in the size of a large cobble but made a great difference within sand-size grades. It also made sense because it allowed a reasonable number of subdivisions of sand sizes that meshed with geologists', engineers', and soil scientists' informal field usage. Using sieves with openings at these intervals, Udden could weigh the proportion of sand retained on each sieve, construct a histogram, or bar diagram, and arrive at an average size and, perhaps as important, he thought, the distribution of sizes about that average (see the figure on page 50).

Size distributions are still determined in much the same way or by use of settling tubes. Armed with modern methods of statistical analysis and computers, the size analyst can quickly find all the vital statistics of sand-size distributions. From these analyses we can distinguish with some confidence the fractions of the sand

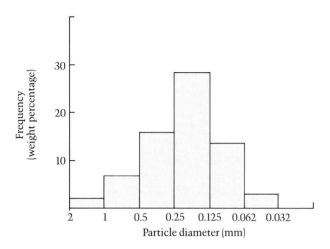

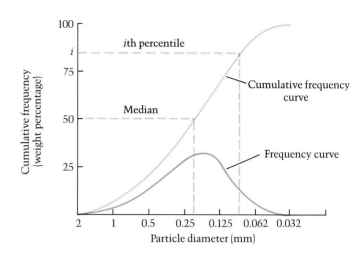

Size distribution of a typical sand. *Left:* A simple histogram showing the relative abundance of grains of various sizes as measured by their diameter. *Right:* The cumulative frequency curve is a plot of cumulative totals as each size-class increment is successively added. The frequency curve can be derived by differentiating the cumulative frequency curve. It can be looked on as the smooth curve that would result from the histogram being broken into infinitely many increments. The median diameter and any percentile can be read from the cumulative curve as shown by dashed lines.

that were transported as saltation load from those that were carried along the bottom as bed load. We also can make astute guesses about the type of current—river, wave, or other—that carried the sand. But even with this array of sophisticated size measures we cannot yet be certain from size analysis alone of which fluid regime deposited a given sand. Most of our deductions still do not stand up to a double-blind test that eliminates the bias of an investigator who, for example, has actually seen the wind blowing the sand. When one person collects a modern sand, gives it to another, who in turn gives it to the analyst, the results turn out to be ambiguous. Our fingerprinting goal remains elusive.

The reasons why our deductions from size analysis fail are not obscure. One is the sand supplied to the current. No matter what the effect of the flow, a current that is supplied with fine and medium sand from the weathering region cannot produce a coarse sand. Some differences in size, then, are attributable to provenance. Another difficulty is that experimenters studying sand-transport mechanics in the laboratory have a hard enough time finding out what happens with a single size or narrow range of sizes. Once a mixture of sizes is introduced into the flume, the analysis gets more difficult. Finally, although a river may look completely different from a beach, the ways in which the currents

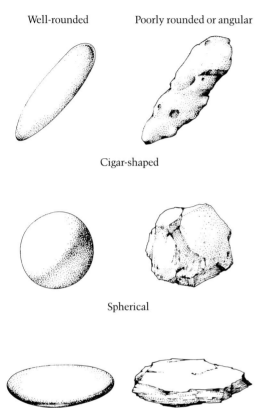

Well-rounded   Poorly rounded or angular

Cigar-shaped

Spherical

Disk-shaped

Overall shapes of sand grains may approach the regular geometries of spheres, cigars, disks, or other common shapes. Roundness is a property of small edges and corners. Shape and roundness are independent, as can be seen from the comparisons in the figure.

of each sort sand grains by size may not be all that different. The jury is still out after a hundred years.

Defining and interpreting grain shape is more of a success story. Almost as soon as geologists learned how to look through a polarizing microscope at sandstones they recognized the similarity of thin, elongate quartz grains to those found in schists. These are metamorphic rocks in which the quartz grains are part of a texture of wavy sheets of cleavage, the tendency of rocks or minerals to split along certain directions, formed by intense directed pressure. In other sandstones almost perfect spherical grains were found. It was easy to jump to the conclusion that these must have formed by wear and abrasion as grains knocked about and hit each other during transport. If sand grains could blast glass bottles to a ground-glass finish, why couldn't they do the same for individual grains, producing a matte, frosted surface that inhabitants of deserts knew was common on dune sands? By now the story was getting complicated. Sedimentologists began to straighten out these different effects by tightening up their definitions of the geometric properties of grains.

About 50 years ago Haakon Wadell, a young graduate student at the University of Chicago tackled the problem anew. Working with William Krumbein and Francis Pettijohn, two pace-setting young sedimentologists who were my mentors as well, Wadell distinguished between shape and roundness. Shape was a larger-scale property by which one could differentiate among spherical, cylindrical, bladelike, or sheetlike grains. By contrast, roundness, or angularity, its inverse, described smoothness or roughness of grains on a much smaller scale that took account of features such as edges and corners. One could find many examples of well-rounded blade- and disk-shaped pebbles on a beach that were far from spherical in shape. In other places more or less equant (all the diameters are roughly equal), subspherical grains are rough and angular. By 1940 the definitions were in place and a variety of methods invented by which to quantitatively measure these properties. Simplest were the scales devised by Wadell in which sphericity—the closeness of a shape to a sphere—scaled from 0 to 1, with 1 a perfect sphere. Roundness similarly went from 0 to 1, with perfect roundness being 1. Now that quantitative measurements of properties became available new attention was given to their geological interpretation.

Definitions of shape and roundness by Wadell. These measurements were made on individual grains from cross sections as seen under a microscope or projected from an enlarger. In practice, silhouettes of many grains of various measured roundnesses were used for visual estimation.

Shape: Ratio of a cross-sectional area of grain to the smallest circumscribing circle

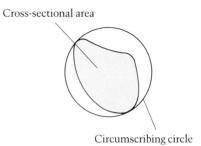

Cross-sectional area

Circumscribing circle

Roundness: Ratio of average radius of curvature of corners to radius of maximum inscribed circle

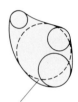

Inscribed circle

At first, roundness measurements were the simplest to interpret. Abrasion did round sand grains and pebbles, and one could construct tumbling barrels in which many kilometers of transport could be mimicked to show just how rapidly it happened. Others went out to the field to hunt for rivers that made good natural experiments to test for the degree of rounding introduced by various degrees of rigor of transport. Bill Krumbein went to southern California in the late 1930s to study gravels dumped by a large flood in San Gabriel Canyon.

On March 2, 1938, one of the largest floods ever recorded wreaked havoc on the Los Angeles area. Several days before, continuous rainfall had saturated the surface of the land. On the evening of March 1, the rainfall increased dramatically. The ground was already saturated, and all of the rain ran into streams and canyons, starting the extraordinary flood. Krumbein saw the potential in this natural experiment to test his ideas on the effects of transport on sediment texture. The debris and confusion left behind by the flood were still there by the time he was able to get there the next summer. (Back in the 1930s, you didn't leave your university responsibilities readily and hop a plane the next day; it was a 40-hour train trip from Chicago to Los Angeles.) He sampled the length of the canyon measuring shape, roundness, and particle sizes, from the sand and gravel that choked the floor of the canyon to the huge boulders that had crumpled houses. As easy as it might have been to be overwhelmed, Krumbein was not distracted by anecdotes and the bizarre particulars of sediment transport by the flood: he followed a carefully devised statistical sampling and measurement program. As one of the pioneers in the use of modern statistics in geology, he knew that the San Gabriel Canyon program was his chance to tease important conclusions from the mass of data. The data, as well as the canyon, were chaotic, reflecting the high, nonsystematic variability of the natural process. But roundness was systematic. The roundness of pebbles decreased so regularly that Krumbein was able to formulate a mathematical function relating distance downstream, the initial roundness, and the final roundness, in which a constant is determined by the specific characteristics of the type of environment.

Getting a general expression for the other textural attributes of the sediment was more difficult because of the heterogeneity of rock types and the many microenvironments within the canyon. The search by Krumbein and others for a perfectly clean natural experiment with simple general results has continued to prove illusory. Few rivers drain completely homogeneous rock terranes and one could never be sure that abrasion effects were not either masked or reinforced by tributaries bringing in different source materials. But enough good, though not perfect, examples were found that matched the admittedly crude experiments to give a coherent picture.

The larger particles, cobbles and pebbles, round fast. Clambering through the jumble of debris in a narrow mountain stream, the geologist can see barely perceptible differences with distance downstream. In only a few kilometers of transport edges and corners are smoothed, and in tens of kilometers a pebble might become well rounded. The rate at which this change takes place varies with the environment. It is much slower in the open, starkly barren rockiness of a desert dry gulch than it is in the humid, rainy, brush-tangled ravines of the pine-forested southern Appalachians.

Sand grains, especially those of quartz, the commonest mineral of sands, are much slower to round. The finer the grain size, the

Sand spits and underwater bars on Cape Cod.

slower the process. This slowness made the results of very long transport perplexing. In simulated transport in tumbling barrels, various experimenters found that thousands of kilometers were needed to produce only moderate rounding on medium-size sand grains. An exhaustive study of Mississippi River sands by Dana Russell was inconclusive, the roundness not changing very much in the 1600 km between Cairo, Illinois, and New Orleans. After joking that he had ruined his eyes by peering down a microscope at those millions of sand grains, Russell decided that there were too many tributaries bringing in sand from diverse geologic terranes to be able to tell whether one of the largest, longest rivers in the world rounded sand grains.

The paradox remained. How could we force nature to give up this little secret? "Well," said one group, "let's go to a beach." There the back and forth motion of the waves induces many kilometers of transport while the sand slowly migrates along a spit of sand. Sedimentologists started to tramp all sorts of beaches to resolve the paradox of rounding. Francis Pettijohn went to a beach spit on the shores of Lake Erie. Krumbein walked his favorite beaches along Lake Michigan. Jack Hough, a student of Krumbein and Pettijohn, chose the shores of Buzzards Bay on Cape Cod. In these and other beaches too, the results suggested that, although beaches were more effective than streams, rounding was still very slow. Yet there remained the sticking point: far too many ancient sandstones, variously deposited by rivers, by wind, or by marine currents, consisted of superbly rounded grains.

The answer to this apparent paradox came over the years as we recognized that many sands are recycled. Sand grains may change slowly, but if they stay in the top 10 km of the Earth's crust, too shallow to be severely affected by the great heat and pressure of deeper zones of the crust, they are almost indestructible. When a loose quartz sand is eroded it leads to the same sand grains being deposited in another place. For the most part, when a quartz sandstone is eroded, it too is broken into its constituent grains. After a sand is deposited, it is buried under layers of subsequent sediment, later uplifted in a mountain belt, then exhumed and eroded to again be transported and deposited. In this cycle, the sand grains that once came together to form the original sand are liberated by a later stage of the cycle. Sand grains have no souls but they are reincarnated. Each cycle of deposition, burial, uplift, and erosion renews the sand grains and rounds each grain a little more. If the average recycling time is about 200 million years—a rough guess—a Devonian sand grain deposited about 380 million years ago might have been recycled 10 times since it was first eroded from a granite about 2.4 billion years ago. That is a history that encompasses more than half the age of the Earth.

Feldspar and other minerals are recycled too but do not last nearly so long as quartz. They are less stable at the Earth's surface than quartz and tend to dissolve or be degraded to silt or clay in one or two cycles. Other grains, volcanic rock fragments for example, may be reincarnated in a different way, by complete transformation during burial. Volcanics are chemically reactive in the somewhat elevated temperatures and pressures of the burial environment. At temperatures of more than 100°C and pressures of hundreds of atmospheres, the volcanic may be transformed completely, metamorphosed to an entirely new set of minerals. To follow recycled material like that takes a geochemical tracer, but that is a story for Chapter 10.

Textures other than size, shape, and roundness may be transitory. The surface frosting common on desert dune sands may not last more than the current cycle, because each burial episode de-

Sand grains derived from the first-cycle weathering of a rock such as granite may be recycled several times. Each cycle may cause additional rounding of grains and the disappearance of feldspar and other unstable minerals.

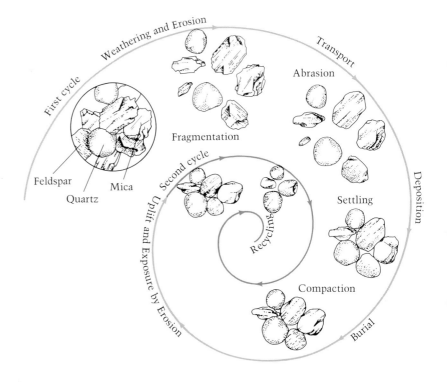

stroys the surface textures of the last as subsurface waters under the elevated temperatures of the interior chemically modify the borders of the grain. Frosting is a texture that is obvious with the naked eye; we need only a low-power binocular microscope to see it on sand grains. But as we magnify more, using the SEM a new world appears. Pits, cleavages, stepped planes, and other geometries difficult to describe in words appear in a bewildering array, sometimes many on a single grain.

The irregularity of geometry of sands seems to follow the rules set out by Benoit Mandelbrot in the 1960s for self-similarity in his early exploration of fractals. As one goes down in scale from outcrop to hand specimen to sand grain to optical microscopic magnifications and then to SEM magnifications, the same crenulated, irregular, bumpy surface continues. Scanning electron microscopy of sand grains is letting us associate different textures with such different environments as windblown, glacial, or beach, but so far we have only poor insight into the mechanisms that produce these textures. One exception is frosting.

Collisions of one grain with another seemed the obvious cause of the presumed microchipping and pocking that would have produced the matte, or frosted, surface. This received wisdom on microballistics went unexamined until Philip Kuenen of the Netherlands, one of the great sedimentologic discoverers of the 1930s and 1940s, looked carefully at these surfaces. An inveterate experimenter, he showed that most of the pits and irregularities of the frosted surface were too small to have been created by the impact of a similarly sized grain. At the same time, he calculated that the momentum of grains small enough to correspond to the pocks was far too small to penetrate the cushion of air surrounding a grain and knock off a piece. Close observation of the grain surface not only allowed him to debunk the conventional wisdom but gave him a better idea. Many of the indentations of the irregular surface reminded Kuenen of etch pits formed by chemical corrosion by a solvent. Racing to his conclusion, he proposed that desert dew—present on some of the dryest terranes—was sufficient over a long time to chemically produce frosting, the way we make frosted glass commercially by using hydrofluoric acid.

When a geologist wants to confirm a hypothesis, the atlas of the world is plucked from the shelf. Though culturally interesting places have no necessary correspondence to particular kinds of geological terranes, geologists frequently turn up in exotic places. Kuenen was a world traveler. In his youth he had been a geologist with the Dutch *Snellius* expedition sent to study the oceans around the volcanic arcs of Indonesia (then the Dutch East Indies). The child of a Dutch father and English mother, he felt some affinity with South Africa, the former colony of his parents' countries. Whatever the complex chain of motivation, Kuenen chose the Kalahari Desert of Africa to be the natural experiment that would confirm his ideas. When the notably frosted sands of the Kalahari were blown into the Zambesi River and transported by the water, the frosting disappeared, and smooth, shiny grains became the rule. Even though quartz is so slightly soluble in water that for practical purposes it is inert, this was circumstantial evidence of dissolution by water, producing frosting on the desert sands and smoothing frosting in river sands. In these two

Typical shapes of minerals forming common sand grains. The shapes are determined by cleavage directions that are reflections of the crystal structure of these silicates.

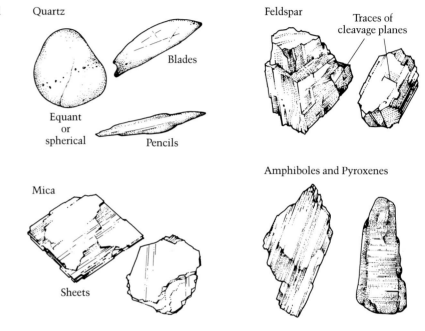

very different environments—a dry, windy desert and a large, rapidly flowing river—water works effectively as a solvent. Current work with the SEM shows that the picture is more complicated, as we might expect from the multiplicity of physical and chemical processes affecting grain surfaces. Nevertheless, a geologist who finds frosted grains in an ancient sandstone is amply justified in making a first guess that they are windblown.

If surface textures quickly respond to changes in environment and roundness slowly increases with transport abrasion, how could geologists explain the variety of grain shapes, from slivers to spheres? With the confusion between shape and roundness cleared away, we could see shape as inherited from the original crystal formed in an igneous or metamorphic rock. Quartz grains could be spherical, blade-shaped, or pencil-shaped, depending on their crystal shapes in the source rock. Feldspars tend to preserve some semblance of their original, more rhombic shapes. Micas are sheetlike; amphiboles and pyroxenes, more pencillike. All these shapes were modified by rounding and fracturing along cleavage planes during weathering and transport. Luckily for the geologist, enough of the original is left in many grains to use shape to confirm mineralogy as a clue to the kind of rocks that were weathered in the source area of the sand.

Our excursion into the details of individual grains has told us much about the results of transport, but even after we analyze all these properties—size, shape, roundness, surface texture, and composition—we still do not have a good way of telling what kind of geological transport agent or environment was the major actor in producing the sand we see. To draw a map of ancient times, we need knowledge that we can get only by moving away from thinking about individual grains to looking at the properties of the bulk—a sandstone as we see it in outcrop.

THE MACROVIEW: BEDDING AND SEDIMENTARY STRUCTURES

The red-orange sandstone cliffs that tower over Zion Canyon spectacularly display lines and planes cutting across sheer vertical faces, especially when the climbing or setting sun lights its craggy walls. This pattern in the rock is not an accident of illumination or of either erosion or coloring imparted by weathering. When you look at the rock using both sides of the brain, integrating the patterns over several scales, two dominant geometries become apparent. The first pattern is the horizontal bedding planes that separate layers of sandstone a few meters apart. The second is the cross-bedding, the bundles of fine layers inclined at about 40° to the bedding. We see both bedding and cross-bedding revealed more clearly in the weathered outcrop than in fresh road cuts, giving the prized scenery of the American Southwest that attracts millions of tourists. Without going to Arizona or New Mexico, you can see these patterns along the highways of southern Wisconsin, along the shores of Lake George, or as I did for several years, on the bottoms of small creek beds in Southern Illinois. Wherever you find sandstone, you are likely to see cross-bedding.

Cross-bedding was known and mapped a hundred years before we began, in the 1950s, to learn how it forms. Before 1950, sedimentologists either ignored it or classified it to death by its many variations in form. The literature, which included unwar-

Cross-bedding in the Canyon de Chelly sandstone, Arizona.

ranted assumptions—speculations unhampered by experiment or detailed observation of modern sands—generally ignored the fluid mechanical approach to the origin of cross-bedding.

The rediscovery of cross-bedding came after modern mapping of sandstone formations confirmed Henry Sorby's hunch in 1859 that the direction of maximum inclination of cross-bedding was the direction of sediment transport. Paul E. Potter, working on his doctoral thesis under Francis Pettijohn in the early 1950s, saw that the Lafayette gravel, a Tertiary age formation, had a coherent set of cross-bedding directions that, when statistically analyzed, revealed the paleocurrents of the ancestral Tennessee River. An explosion of paleocurrent mapping followed, initially dominated by Pettijohn and his students. Soon after finishing the Lafayette gravel, Potter and I joined in extending the mapping of paleocurrents over a larger region than had ever been covered before.

The task was formidable: to map cross-bedding in a particular sandstone formation over the states of Illinois, Indiana, Ohio, Kentucky, and Tennessee and over parts of Pennsylvania and West Virginia. If we had mapped as geologists are wont to do, walking over parts of the whole territory, noting a multitude of rock characteristics and slowly constructing a map bit by bit, it would have taken years. We were not only impatient but convinced that we could do it in a matter of weeks using statistical sampling methods. Our plan was simple: use a hierarchical sampling design and ignore all the other interesting geology; just measure the cross-bedding. We measured the direction of two cross-bedded units at each of two outcrops in each of two square-mile sections in each of 36 square-mile land survey townships along the entire outcrop belt of sandstone. We kept up a frenetic pace, driving along the road until we saw an outcrop, jumping out of the car, with one measuring and the other recording (remember, this was in the days before tape recorders), then, as soon as we were through, shouting "Let's go!" and scrambling back to the car. We even begrudged the time taken for a cold drink in the middle of a hot, dusty afternoon.

In a matter of weeks we accumulated the data for a paleogeographic reconstruction of a great river system flowing over the eastern third of the North American continent during the early Pennsylvanian, a time when most of the coal deposits of North America were formed. We were not the only ones. Francis Pettijohn, who was the first to recognize the potential of cross-bedding at that time, mapped most of the sandstones of the central and southern Appalachian region with his students and synthesized a model for the evolution of sandstones of a major

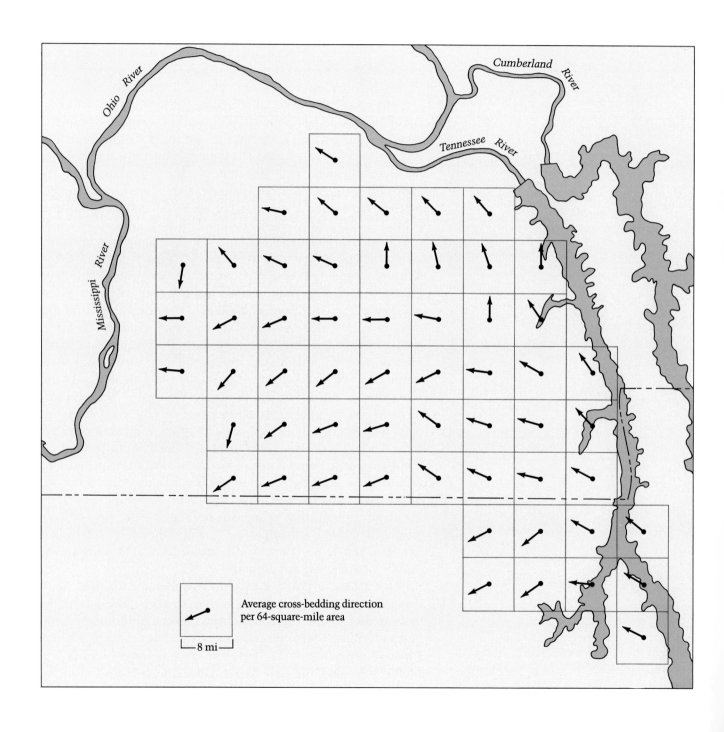

Average cross-bedding direction
per 64-square-mile area

8 mi

Opposite: Cross-bedding in the Lafayette formation, depicted as a moving average for a geographic grid. The trend of these arrows defines the general paleocurrents for the region of western Kentucky and Tennessee. *Right:* Paleocurrent directions based on cross-bedding in the basal Pennsylvanian sandstones of part of the northeastern United States. General directions correspond to areas distinguished by the presence or absence of metamorphic quartz pebbles.

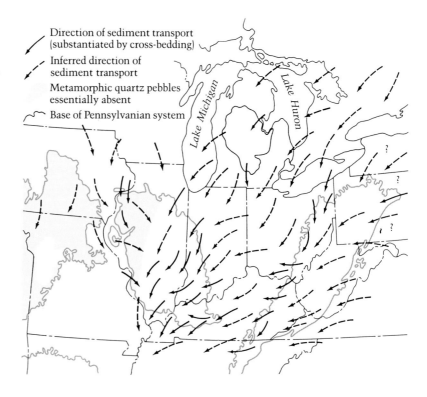

mountain belt through the several hundred years of its development.

This spurt of activity on cross-bedding inevitably led to questions about the fluid mechanics responsible for such an array of different conformations in sandstones formed in so many different environments. Flume experimenters turned their attention to a new problem and soon discovered what geologists had known for some time, that ripples and dunes were cross-bedded as an integral part of their formation. The mechanism could be traced to another aspect of some fluid flows, flow separation.

If you have ever stood in shallow water in a rapidly flowing stream and felt the tug on your legs, you are familiar with flow separation. A large obstacle in the path of a rapid flow, which for our purposes is best illustrated by a ripple or dune on a sandy stream bottom, will cause the boundary layer to suddenly expand to the point where the flow separates into two regions, the main

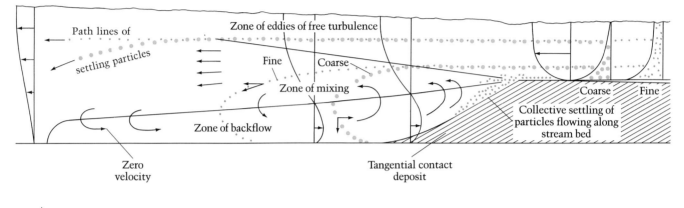

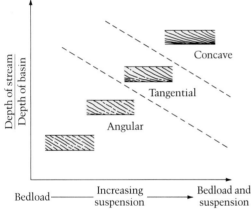

Top: Trajectories of grains and flow separation at the front of an advancing cross-bedded unit. Velocity profiles are shown at verticle bars and sediment concentrations at far right. *Bottom:* Forms of cross-bedding in relation to depth of water and amount of suspended grains, which is dependent on velocity.

flow and a reverse eddy. Downstream from this detachment point, the flow gradually recovers; the main flow expands, at the expense of the reverse eddy, to an attachment point where the backflow disappears and the flow returns to normal. As a result of flow separation, a distinctive set of grain motions preserves the wave form of the ripple as it migrates downstream.

At flow velocities at which ripples can form, sand grains are rolling and saltating in low jumps along the bottom. As they follow trajectories up the upstream slope of the ripple, they reach the crest. Here they meet the point of detachment and one of several paths can follow. A small, high-jumping grain will jump off the crest and find itself in the main forward flow, to be carried some distance downstream before landing beyond the attachment point. It has not even been involved in the flow separation. A medium grain has no such luck. It does not jump as high and finds itself quickly dropping into the reverse eddy, where its former forward motion is countered by the backflow, which carries it toward the foot of the ripple. A third grain, too heavy to saltate rolls over the crest and lodges on the upper part of the steeply angled downstream, or slip, face. After many grains fall there, the buildup becomes gravitationally unstable and a mass of grains cascades down in a minilandslide. The new angle of repose of that slope is the gravitationally stable one, identical with the angle of

Grain trajectories at the edge of an advancing cross-bedded unit.

cross-bedding we see in the field. In this way the wave form of the ripple is preserved as it slowly moves downstream by the aggregation of thousands of individual grain trajectories. The flow separation forms the ripples and is itself produced by the ripple in a feedback loop.

As ripples and dunes travel downstream and build up a layer of sand, they produce a cross-bedded layer about as thick as the height of the wave, with an angle controlled by flow velocity, grain size, and thickness or height of the flow. Either a faster flow or a smaller grain size will result in longer trajectories and a lower angle of the slip face and thus the cross-bedding. A thicker flow will produce the same result. Slower flows and larger grain sizes produce higher angles. High-angle cross-bedding had for a long time mistakenly been called torrential. As it turned out, the word *tranquil* would have corresponded better to the kinds of flow responsible. As is often the case, intuition is a better guide when solidly based on theory and experiment than when based on guesswork.

Once we started to understand how ripples and dunes are cross-bedded it was natural to move to the differences in flow that would account for different bedforms: small ripples, large ripples, the larger forms we call dunes, and ripples superimposed on dunes. We already knew the results of a simple, now-famous experiment performed in 1914 by G. K. Gilbert, a great geologist of

Ripples on a sandy beach exposed at low tide.

his time, equally at home in the field and laboratory. Starting with a very-low-velocity flow over a sand bed in a flume, he gradually increased the velocity. At the lowest velocities only the smallest grains rolled along a smooth bed. As velocities increased, more and more grains were moving but the bed was still smooth. Suddenly, with the next increment of velocity, small ripples furrowed the bed as random clumpings of grains and an instability of the bed caused a change in bed state. As velocity further increased, the ripples grew and migrated downstream at a faster rate. By now, the water was flowing at a good clip. The next velocity increase formed dunes and then ripples migrating over the backs of the dunes. This progression stopped with the next increase in velocity, when all dunes and ripples were suddenly wiped out and a new plane bed was formed. This plane bed was vastly different from the initial one. Virtually the entire surface of the bed was in motion and the saltation load was large. When Gilbert turned up the velocity still more, the whole bed went into motion; both fluid and bed deformed into standing waves that slowly migrated upstream, occasionally becoming unstable and collapsing in a turbulent crash.

Modern experiments have reproduced this sequence many times and given an explanation fuller than Gilbert's for the variations in bedform, the type of flow, and the roles of velocity, depth of flow, and grain size, as shown in the figure on page 67. John Southard and his students at the Massachusetts Institute of Technology (M.I.T.) have explored this three-dimensional space in recent years. Southard, interested in field sedimentology as an undergraduate, saw the promise in the application of fluid mechanics to sedimentology. As a graduate student working in my laboratory at Harvard, he took a full load of courses in fluid mechanics and did an experimental fluid mechanical thesis. Equipping his laboratory at M.I.T. with elegantly instrumented flumes, he varied all the parameters and found that the sequence of bedforms Gilbert had observed was not universal. For some small grain sizes, for example, no dunes formed; the bed went directly to the higher-velocity plane bed from a rippled state.

Whatever the particular sequence, the transfer of this experimental information to the field was not long in coming. We can view Gilbert's experiment as a model of a river going into flood. Direct natural observation of the bottom of a river in flood is not a

Bedforms produced for various combinations of velocity and flow depth with sand sizes of 0.45 to 0.55 mm. Each triangle represents one experiment at fixed velocity and depth.

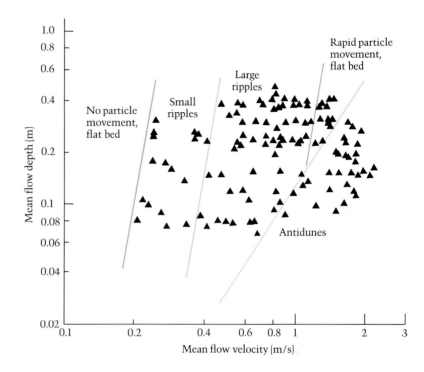

recommended undertaking. The next best thing is to cut trenches at low water across and along river bars known to have formed or migrated during floods. There geologists have found an array of cross-bedding types that they could correlate with specific flow conditions as the river went into flood and then receded. But that is part of the next chapter's story, how field sedimentologists learned to recognize the imprint of sedimentary environments on sandstone bedding sequences and thereby distinguish river deposits from windblown and beach deposits.

CHAPTER FOUR

Facies detectives

I enjoy taking a physicist to a rock outcrop. Physicists make you look with new eyes at familiar—hardly noticed—aspects of the rock, like its stratification. "Starting from first principles," I ask, "how would you be able to tell if this outcrop was once an ancient stream deposit? I can infer it from geology, but can you derive it from mechanics?" A smile and a shake of the head is the response. The question is unfair, for a direct jump from the newtonian mechanics that govern sediment transport to the complexities of the natural rock is virtually impossible. It is hard enough to figure out the dynamics in a flume, given the many variables, the nonlinearity of flow turbulence, and the uneven, stochastic motion of the grains. In the rock all you can see is the frozen snapshot of the bed that was once deposited from that unknown flow. Which of the many aspects of the rock that you can describe should you pick to reconstruct those long-lost dynamics? Grain texture? Bed thickness? Bulk density? Out of dozens of available properties, cross-bedding seems like a good place to start.

Given the grain size that we can observe, we ought to be able to reconstruct a flow regime from the cross-bedding—but not without ambiguity. We can work backward to deduce various possible pairs of flow velocity and depth of water from the measured cross-bedding angle, the thickness of the cross-bedded unit, and the grain size, but we cannot get a unique solution. We also have to consider the vertically changing geometry of the cross-bedding, which might include large sets of tabular low-angle cross-beds overlain by thinner, more troughlike units of finer grain size and steeper inclination. To make sense out of these variations, we have to plug ourselves into the lore of river beds.

Carrying shovels and spades, and sometimes driving a bulldozer over a dry river bed, field sedimentologists working on modern river deposits might be mistaken for construction workers.

Weathered cross-bedding in the Navajo Sandstone of Zion Canyon.

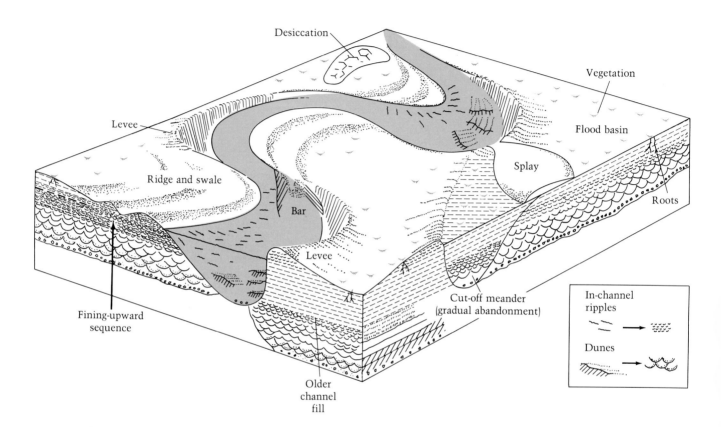

The fluvial environment, showing typical relationships among channel floor and bar, levee, and floodplain deposits. Internal structures of cross-bedding and ripples are correlated with these deposits.

We need these tools to dig trenches across sand bars and the floors of channels that let us see cross sections of cross-bedding, ripples, and other structures. We can then compare the deposits accessible at low-water periods with the dynamics of river flow at higher-water times; it is clear from the river bed that nothing much happens at very-low-water intervals. At high-water periods, we can measure the flow with current meters or pressure transducers; we can then sense the configuration of the bed by echo sounding, but we cannot detect the internal structure of the forms at the bottom. The trick is to correlate the internal structures found at low water with the flow characteristics at high water and set all this against a matrix of many seemingly unrelated characteristics of the cross-bedded rock that might be additional clues to its origin.

The facies, or overall aspect, of these properties of a sequence of beds differentiates these beds from bed sequences that have

The fining-upward cycle, the gradual change from coarse gravel at the base to fine silt and mud at the top, produced by the normal operation of a meandering river. Cross-bedding and ripplemark change upward with grain size.

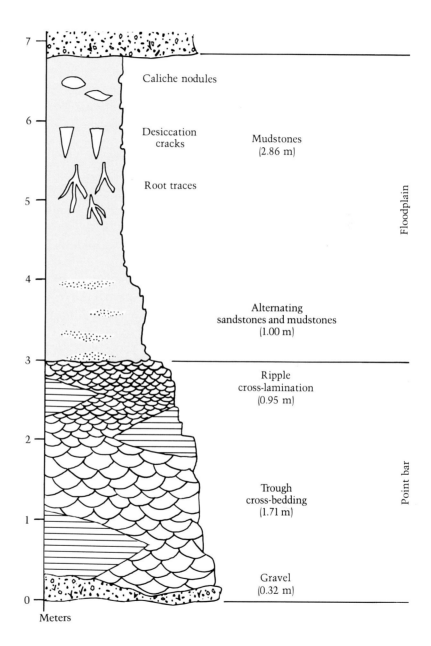

somewhat different sets of properties. In the early days of mapping cross-bedding, when Paul Potter and I were in the field in eastern Kentucky, we differentiated between a parallel-bedded and cross-bedded facies without making any further designation, even though we both could have listed many other characteristics that lay in the backs of our minds. Because a full description would include so many properties with complex parameters, sedimentologists use this kind of shorthand. If we are prepared to go out on a limb we can label it an "alluvial channel facies" or a "delta plain facies." We do not use these terms as purely descriptive measures. Instead, we frequently use the depositional environment as a designation because it is the best means we have, even though it implies the conclusion we are trying to reach. For example, we might study the cross-beds and other properties of a particular sandstone facies and find them similar to the alluvial channel facies of a modern river. Instead of naming the sandstone facies by some purely descriptive term, such as the trough cross-bedding facies, we use the inferred origin and name it for the comparable modern facies, the alluvial channel sandstone facies. As central to this science as the idea of environment is, it is by no means a simple concept.

SEDIMENTARY ENVIRONMENTS

Typical of all coiners of terms they think of as their own, sedimentologists were a little peeved when in the late 1960s, *environment* seemed to be on everyone's tongue. If precise usage is the standard, we can ignore this gripe because the sedimentologists are not so precise either. What, for example, do they specify by the term *sedimentary environment*? We could lump together all turbulent sedimentary environments no matter where they are found. We could define environments by their chemical characteristics, such as the normal salinity of marine environments. Most such terminological schemes lack any coherent sediment facies that would correspond to them. But such facies do exist for environments defined by reference to a group of properties that in shorthand we call a "river" or a "beach."

A sedimentary environment, then, is broadly construed as a kind of place, of broad dimensions, whether at the surface of the land or at the bottom of the sea, in which a group of sediment-forming processes produce a recognizable corresponding sedimentary facies that can be mapped. The river, or alluvial, environment includes not only the channel in which the water flows most of the time, but the floodplain beyond the channel banks,

Left: The braided Chitina River, Alaska. *Right:* Meanders in a river flowing into a tidal salt marsh.

where floodwaters deposit silt and mud only once in many years. One can include the valley borders and upper slopes. The subdivisions form subfacies, but the dynamics of the river ties together the entire aggregation so that we can recognize the facies by the association of subfacies. Not all rivers are alike; for example, meandering streams have one sinuous channel, and braided streams have dozens of channels that anastomose, or divide and reunite, along a general channel locus.

As shown in the table on page 74, the list of sedimentary environments is not over long. For example, on the land surface a lake is the only major watery environment other than a river; eolian, or wind-dominated, environments might be found in the dune belts along rivers in arid terrains, at the backs of many beaches, or in the desert. The desert environment itself is somewhat differently defined as a place dominated by dryness, where streams run rarely and wind blows most of the time. Nonwatery but icy environments include the glacial areas of alpine heights or the vast ice fields of Antarctica. Undersea environments, subdivided by depth of water or distance from shore, include beaches and nearshore

Sedimentary environments of sand

Environment	Description
Continental	
Alluvial fan	Aprons of coarse river-laid sediment at the edges of valleys bordering steep mountains
Alluvial plain	Channel and floodplain deposits of braided- and meandering-river valleys
Desert, or eolian	Sand-dune fields of deserts; semi-arid, broad, sandy river valleys; and coastal dune belts related to beaches
Lake	Similar to shoreline deposits of deltas, but smaller in scale
Glacial	Till and outwash plains of continental and valley glaciers
Coastal	
Delta	Shoreline deposits dropped by rivers at their mouths, modified by waves and tides
Beach and bar	Sand, including submerged or island sand bars offshore, distributed parallel to the shoreline by waves and tides
Continental border	
Continental shelf	Sand ribbons, sheets, and dunes on shallow (<150 m), broad (>50 km) continental margins
Continental slope and rise	Slumps and slides, submarine fans, and turbidity-current deposits
Pelagic (open ocean)	
Abyssal plain	Thin sheets of fine sand, silt, and mud deposited by turbidity currents running out from continental margins

bars, offshore bars and shallow-marine continental shelves, the deeper continental slopes and rises, deep-ocean trenches, and abyssal plains, extensive flat areas of the deep oceans. Thanks to an immense amount of field observation in the 1960s and 1970s of modern sediments, their transport agents, and comparable ancient rocks, we can recognize the sediments of all these environments.

Satellite photograph of one arm of the Ganges River entering the head of the Ganges delta, Dacca, Bangladesh.

Midocean ridges—some of the most fascinating undersea envi-ronments—contain too little sediment to be included in most lists of sedimentary environments. Here, in the midst of active volcanism and rifting between newly created plates, sedimentary processes play a minor role. By the time any sediment slowly accumulates in these faraway parts of the oceans, the plates have moved away from the ridge. As they move away and calm down, they become more interesting to the sedimentologist: the cal-cium-carbonate shells of foraminifera and fine clay then settle and round the once rugged underwater topography with a mantle of soft sediment. In contrast, the island arc and trench associated with a subduction zone, which is another active plate-tectonic setting, includes sediment of great interest. However, the sedi-ments of these tectonic settings are more profitably analyzed

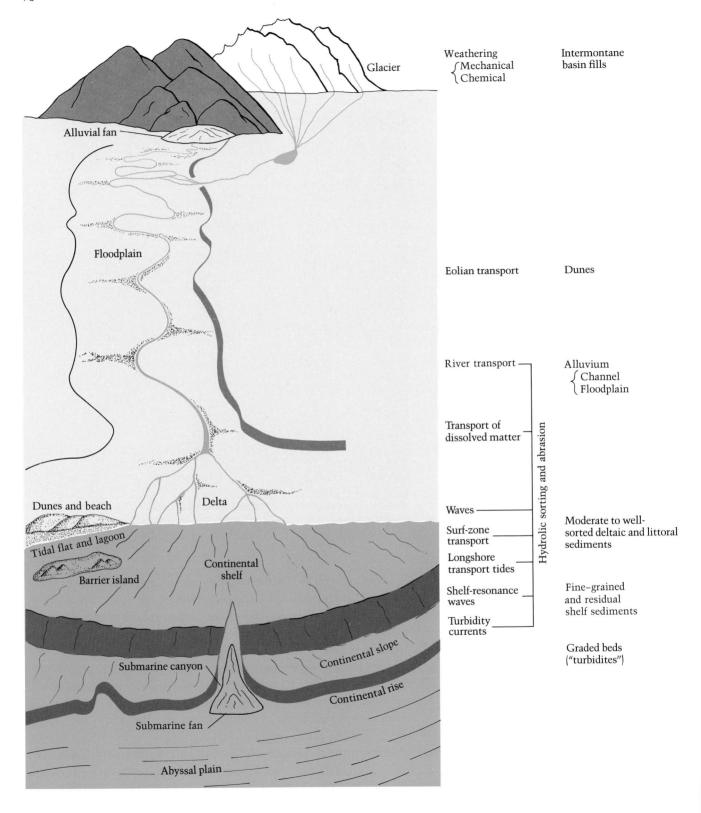

Glacier

Weathering { Mechanical / Chemical

Intermontane basin fills

Alluvial fan

Floodplain

Eolian transport

Dunes

River transport

Alluvium { Channel / Floodplain

Transport of dissolved matter

Dunes and beach

Delta

Tidal flat and lagoon

Barrier island

Continental shelf

Waves

Surf-zone transport

Longshore transport tides

Shelf-resonance waves

Turbidity currents

Hydrolic sorting and abrasion

Moderate to well-sorted deltaic and littoral sediments

Fine–grained and residual shelf sediments

Submarine canyon

Continental slope

Continental rise

Submarine fan

Graded beds ("turbidites")

Abyssal plain

Opposite: Sedimentary environments, from mountains to deep ocean, sediment-forming processes, and sediment types.

when reduced to the standard sets of land and sea environments—primarily because we can infer those kinds of environments from the kinds of sedimentary facies we can map from deep-sea drilling data and rock terranes formed in these settings.

Sedimentologists achieved the goal of environmental recognition by using three kinds of information—modern sediment study, laboratory experiment, and mapping ancient sedimentary rocks. Modern sediment study has given us the sediment-transport dynamics linking the transport agent with the sediment deposited. Experiment has taught us how to use simplified models to analyze the major forces and concentrate on the most important parameters of both flow and deposit. The mapping of ancient rocks contributes the third dimension, the vertical change in a rock sequence that gives us the time-dependent variation in the system. The facies detectives who have worked this out started with the most accessible and therefore one of the best-known environments of the land surface, the river. In the sections below we will contrast rivers with deserts, less well known to many of us, but fast becoming familiar to modern sedimentologists.

THE ALLUVIAL ENVIRONMENT

The Romans used the word *alluvium* for the sand, silt, and mud a flooded river leaves behind when it recedes to its channel. Long before them, the Egyptians had a general idea of river dynamics and its critical role in their agricultural economy along the Nile. Every year the river swelled in the spring, usually staying within its natural levees, the raised banks of the channel. Every few years the waters rose high enough to spill over onto the broad floodplain bordering the channel, laying down a thin layer of silt and clay that could be cultivated for the next crop. At longer intervals of several decades much larger floods had less happy consequences. The channel might migrate from its former position and carve out a new one, leaving behind a useless coarse-sand deposit that only interfered with farming. Even if it returned to its former channel, a sandy splay covering acres would have been formed where the river broke through its banks and flowed out at high velocity.

A geologist today can account for the deposits by explaining how a flood affects velocity, grain size, and water depth. As a typical river goes into flood, the increased discharge, or amount of water flowing per unit of time, is accompanied by an increase in velocity as the river level rises. Increases in velocity and depth of water are in turn accompanied by an increase in the grain size that the flow can carry; thus, the river starts to erode its sandy bed, transporting much more and much coarser sand in suspension. Soon the river completely fills its banks and then breaks through. As the water pours out of the channel onto the floodplain, it leaves the confinement of the channel banks and spreads out, so that its current velocity quickly decays. In just a short distance from the breakout, the current has lost its velocity, decreased its height, and dropped much of its sandy suspension. Over much of the floodplain the current flows much more slowly than in the channel and transports only fine-grain sediment, silt, and clay. As the discharge lessens, the velocity drops, the water level falls, the river returns to its channel, and the waters of the floodplain become still. The muddy waters clear as the fine clay settles to the bottom; finally they evaporate or infiltrate the ground. The whole operation has taken only a few days.

During this time the river deposits are variously rippled and cross-bedded, depending on the distribution of flow regimes in the channel and floodplain. Channel, splay, and floodplain deposits fit together in a pattern determined by the size of the river, its floodplain, its valley width, and the magnitude of the flood. Most of this work of the river is done not by the many small floods that recur every few years or by the giant flood that occurs only once every several hundred years but by the moderately infrequent, large flood that recurs every few decades. Yet we cannot assign all the erosional and depositional work of a river with certainty to floods, because the day by day migration of a meandering river does its own share of the work.

At its ordinary, below-flood-level stages, the river flows with its highest velocity near the outside bend of a meander loop, and there it erodes the bank. On the opposite bank, at the inside of the bend, the velocity is slowest as the river impinges on the point bar, an accumulation of sand dropped as the velocity decreases toward the bank. The combination of erosion and deposition has an unexpected effect. It moves the river channel toward the out-

Top: Velocity profiles along the bends of a meandering river. The lateral motion at the top and bottom of the channel, together with the forward flow, indicates a helical flow. *Bottom:* Changes in the profile of a meandering channel over a few years, showing migration to the outside (concave bank) of the bend.

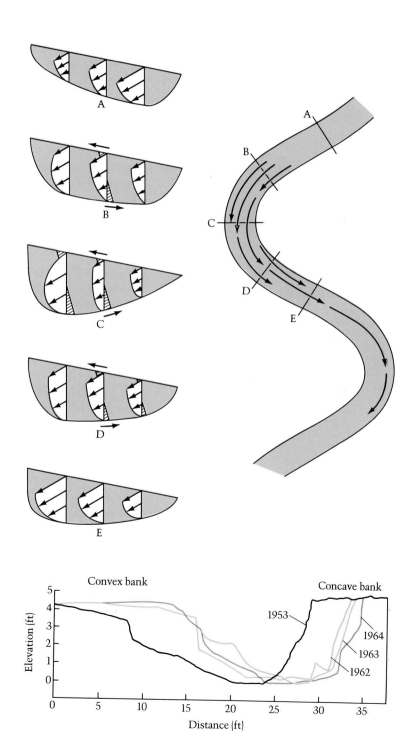

The Old Red Sandstone of England (Pembrokeshire).

side of the bend. As the meanders swing back and forth across a broad floodplain, the channel's deposits of gravels and coarse cross-bedded sands are overlain by point-bar sediments, medium- and fine-grained sands, with trough cross-bedding, which are in turn overlain by silts and muds of the low, marshy borders of the point bar. This progression gives rise to an ordered sequence, from coarser-grained sand and larger-dimension cross-bedding at the base to fine-grained sand and rippled beds at the top of the sandy part of the sequence of beds, followed by silty and muddy beds. In 1964, J. R. L. Allen of Reading University in England first set out to analyze this alluvial fining-upward cycle, now recognized as characteristic of alluvial deposits throughout the world. Allen, a leader in the application of fluid mechanics of sedimentation, saw many such cycles in the Old Red Sandstone of Britain, one of the classic sandstones that figured large in the history of geology. Though the Old Red Sandstone had been studied and restudied by generations of geologists, it was not until Allen perceived the vertical succession of beds, with its fining-upwards property, as an important aspect of the rock that the orderliness of the sequence became apparent.

Later, when Allen visited New York and saw sandstones in the Catskill Mountains of the same age (Devonian period, about 350 million years ago) and facies as the Old Red, the geological world had changed. Plate-tectonic theory was now almost upon us. With the resurgence of interest in continental drift, it was a short step to use new reconstructions of the Devonian continents and envision a single, large "Catskill–Old Red" river system. This ancient river network flowed from a great high-mountain chain created by the assembly of North America, Eurasia, and Africa to form the supercontinent, Pangaea. What a jump in scale! Fluid mechanics and the analysis of small-scale motions of sand grains had been united with the largest motions of the Earth's interior and the continents. But at intermediate scales, that of individual river channels, more research was needed.

The exploration of river deposits had not proceded very far when the importance of different river channel types and their deposits became apparent. Geologists of the nineteenth century had left the low mountains and temperate climates of eastern North America and the Alps and humid climates of western Europe and explored the high mountains and arid regions of western

The distribution of the Catskill–Old Red Sandstone facies in and bordering the Mid-Paleozoic mountains formed by the collision of Europe and North America. The reconstruction of the former position of the continents shows this facies, now separated by the Atlantic Ocean, was formed along a single, connected mountain belt.

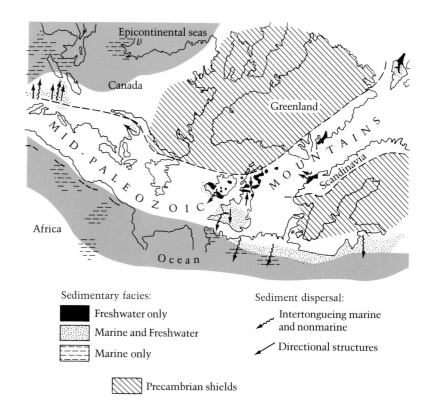

Sedimentary facies:

■ Freshwater only

▨ Marine and Freshwater

⊟ Marine only

▨ Precambrian shields

Sediment dispersal:

⟋ Intertongueing marine and nonmarine

⟋ Directional structures

North America, the Middle East, and Asia. There, finding new patterns of channels and floodplain deposits, they wrote descriptions that were invaluable to later geologists. Two of the patterns, alluvial fans and braided streams, are now known to be as important kinds of river forms as the more familiar single-channel meandering river. The deposition of alluvial fans, large cones of coarse debris dumped by mountain streams as they came down steep mountain fronts, followed rules different from the ones governing the gently flowing rivers of humid plains regions. At the bottom of the mountain front a stream leaves a narrow valley confined by steep walls and suddenly decreases its velocity as it opens onto a broad intermontane valley floor. Gravels, coarse-grain sands, and muddy mixtures of all sizes are spread over the fan as the stream loses its power to transport material. The mud-rich gravels and sands are dropped by debris flows: thick, viscous

Alluvial fan, Death Valley.

flows form as water entrains enormous quantities of previously deposited loose materials of all sizes. Settling velocities are low and virtually all sizes stay in suspension in this high-viscosity, mud-laden flow. Debris flows help to identify alluvial fans, which in turn aid in mapping old mountain fronts, themselves long gone. Put another way, recognizing sedimentary environments tells us about paleogeography, which tells us of plate tectonics.

Differentiating meandering from braided streams is more difficult. We are hampered by lack of a good theory for the two types of channel. The braided habit seems to be more characteristic of flows with both large sediment loads and banks that erode too easily to confine the flow to a single channel. However, many exceptions to these generalizations make us still unclear as to why a river will be braided along some stretches and meander along others. Unfortunately, the sizes of these forms are too great to permit the design of reasonable scaled-down experiments. With these uncertainties, trying to identify ancient braided river deposits depends only on a correlation of channel forms with sed-

Melting snow provides water for braided-stream flow on a sand dune, Coral Dunes, Utah.

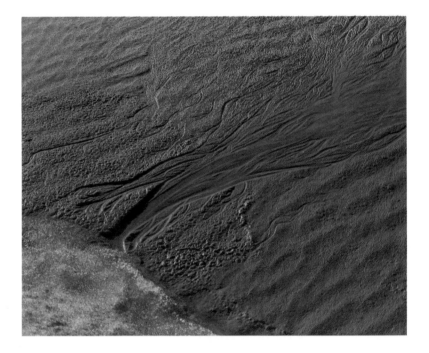

imentary facies and is less satisfactory than a theory based more strongly on fluid dynamics.

We live in towns and cities along rivers, swim in them, use them for water supplies, and see them through the year as they go from flood stages to near dryness, when we can walk over the ripples and bars on the bottom—the relics of channel flow. We are less familiar with large sand dunes, and still less with sandy desert environments and large fields of dunes.

DUNES AND DESERTS

Belts of sand dunes border many of the world's sandy beaches and some arid-region rivers, but the place to see huge dune fields is the desert. The closest most of us come to these seas of sand is in a theater seat, watching a movie filmed on location on a sandy desert. But for the nomadic peoples who have crossed and re-crossed the major deserts of the world for millenia, leaving little in the way of written records, the desert is home. Since medieval

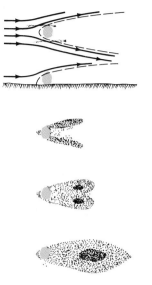

The formation of a small sand dune behind an obstacle, according to Bagnold. *Top:* The wind flow is deflected by the obstacle and leads to sheltered regions. *Bottom:* Successive growth stages of the dune.

times many Moslem peoples of the Near East have recorded observations of the desert and named its land forms in their languages, as have the East Asians who traveled the Gobi.

Modern observations based on a thorough knowledge of mechanics waited for Brigadier Ralph A. Bagnold of the British Army. Assigned to the Libyan and Saharan deserts in 1925, Bagnold mapped dunes, observed the effects of the wind on sand, and brought to his efforts a modern geological and physical viewpoint, as the title of his book suggests: *The Physics of Blown Sand and Desert Dunes.* For example, he explained some of his observations—and frustrating experiences—by the mechanism of dune formation. His Land Rover was able to roar up the hard pavement of the windward side of a dune but instantly sank to its hubcaps in the loosely packed sand of the slip-off slope on the lee side. He related the dense packing of the sand grains on the windward slope to the accretion by the impact of saltating grains and the loose packing of the lee slope to the open structure produced by cascades of sand forming the slip face below the crest of the dune. In howling sandstorms or in calmer but steady blows, he could see how a small obstacle such as a few pebbles could be the start of sand piling up and eventually forming a small dune. In both examples he worked from individual particle motions to the properties of the larger form. For large-scale structures such as dunes, he started to make sense of the many different forms by correlating them with wind velocities and directions.

We can see the relation of form to wind best by looking at the smallest forms, the ripples covering the surfaces of many dunes. Flatter and more regular than underwater ripples, they move rapidly, sometimes at rates of centimeters per minute, in response to the local wind. The crests of the ripples sweep around the dune, from the regular march up the gentle windward slope to the curving patterns around the slopes to the swales between dunes. Plane beds form at higher velocities, as they do underwater. Dunes have an enormous range of sizes, from less than 1 m to more than 100 m high. Ripples form transverse to the wind, but the dune itself may be either transverse or longitudinal. That difference in relation of the dune's elongate dimension to wind direction is related in turn to the uniformity of the wind direction: the more constant the wind's direction, the greater is the dune's tendency to be transverse. In the hierarchy of desert sand forms, ripples

 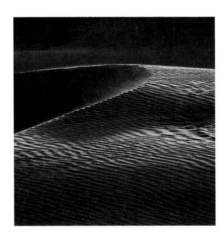 

Ripples on dunes. *Left:* Typical pattern on a windward slope. *Center:* Ripples at the crest of a barchan, or horn-shaped, dune. *Right:* Sand cascades in a pattern of grooves and ridges down the slip, or lee, face of a dune.

climb over dunes, and dunes are small part of *draas*, the Arabic word for great mountains of sand, that may be 0.5 km high—but that is not the end of the hierarchy. Draas and dunes are components of ergs, immense sand seas covering up to 500,000 km^2—twice the size of Nevada.

Sand seas which are few in the American Southwest but common in the Middle East, have formed in large depressions with a ready supply of sand. Typically, abundant sand is supplied from eroding sandstones deposited in an earlier age. The wind pattern is sufficient to blow the sand in but not to allow much of it to exit. Ergs are seemingly endless expanses of sands. Strips of longitudinal dunes cover parts of them; draas take over in other parts.

How mobile are these sand seas? We now suspect that larger dunes, draas, and ergs, now stationary, may be relics from a climatic regime thousands of years ago that had a different spectrum of wind power and direction that moved these enormous quantities of sand. If so, the drastic changes in the Earth's climate that accompanied glacial advances are evidenced by these relics of once-mobile ergs.

The external forms of desert sands appeal to photographers, but geologists take equal interest in the internal structure: the crossbedding. It is entombed with the deposit, waiting for erosion to uncover it and the sedimentologist to decipher its meaning. From the heirarchy of small-scale ripple cross-bedding to giant crossbedded units more than 100 m thick, we can extricate patterns

Cross-bedding exposed in a weathering dune,
Death Valley.

of sand deposition in former deserts. We can map the varying directions of cross-bedding to get paleowind maps analogous to the paleocurrent maps of river deposits that we discussed in Chapter 3. Paleowinds lead us to continental drift, for belts of trade winds and prevailing westerlies are determined by latitude.

In the 1950s, during the early days of paleomagnetic studies, a few geophysicists—one leader was Keith Runcorn of Newcastle University in England—were correlating paleowinds with former latitudes determined by apparent former positions of the magnetic poles, well before anyone had thought of plate tectonics. Geophysicists were not sure about the mechanism of polar wandering, and sedimentologists were reserved, to say the least, about recognizing ancient wind deposits identified by these correlational methods. This is one of the classic stories in which the scoffers came to eat their words. Plate tectonics made continental drift and polar wandering respectable, and sedimentologists learned enough about eolian, or wind-deposited, sands to recognize them with some assurance.

Windblown sand, much of it eroded from older sandstone formations (background), Monument Valley, Arizona.

Today we regularly consult maps of paleolatitudes in reconstructing the past, for example, in the sandstones of the dinosaur age in the American Southwest. In many parts of the Colorado Plateau, especially in some of the national parks of Utah, Colorado, New Mexico, and Arizona, erosion has carved striking forms in sandstones of Permian through Jurassic ages. Monument Valley, Zion Canyon, Canyon de Chelly, and Canyonlands all owe their popularity to the sculpture of erosion. The array of forms results from the interaction of the sandstones' composition and texture with the agents of erosion, rain and wind. A vast continental interior desert formed over the Southwest as the last Paleozoic seas left the continent during the final stages of the assembly of the supercontinent, Pangaea, about 250 million years ago. Edwin McKee, a geologist working for the U.S. Geological Survey, spent a lifetime of fieldwork laying the foundation for this interpretation. Eddie McKee influenced me from the start: when I was ten my family toured the Grand Canyon, and my older brother had brought McKee's popular account of its geology. Although overwhelmed by the beauty of the canyon, I took the guide to the back seat of the car and read it while my family saw the sights. Burying my nose in this book had much to do with my later decision to be a geologist.

Much as he loved the Canyon, McKee wanted field observations from other areas and traveled the world looking at dune sands and deserts. Like his forebear G. K. Gilbert, he experimented with flumes and wind tunnels to simulate what he saw in the field. The major evidence for his conclusion that many of these sands were eolian was the size of the cross-bedded units: layers many tens of meters thick that he thought could come only from windblown deposits. As we all learned more about ripples, dunes, and ergs, his conclusions were confirmed.

Objections taken earlier to this interpretation became part of the reconstruction. In places, some of the sandstones definitely seemed to be river deposited, not eolian. But this argument against McKee's theory evaporated as sedimentologists began to recognize that desert depositional regimes typically include much alluvial sand deposited from the infrequently running rivers fed by rare rainstorms. Dinosaur tracks could be preserved along the banks of the very few constantly flowing major rivers, but the banks of the hot, dry beds of intermittent streams would

Dinosaur tracks in an ancient sandstone near Cameron, Arizona.

be bordered by extensive dune fields. Although marine geologists discovered that high sand waves on the continental shelf also could have large-scale cross-bedding, the Colorado Plateau sandstones held no marine fossils, only some tracks and trails of land animals, and only rarely preserved vegetation.

As we map more sandstones in different parts of the Colorado Plateau, we get a clearer idea of the evolution of this desert. As the last seas of the Permian period withdrew, former coastal plains merged with broad river valleys and upland plains and hills. As the ocean receded farther, to hundreds and then thousands of kilometers away, the climate became dryer. Vegetation then became scarcer and the soil more vulnerable to wind erosion. Sandstones of earlier ages weathered and supplied recycled sand to the wind. Blowing sand and dust covered the dreary lands except near the vegetated borders of the few permanent streams. At the same time, in central Colorado a mountain range was uplifted. These ancestral Rockies, as geologists like to call them, eroded to give coarse gravels and more sand. In just a few tens of millions of years the broad region had been converted to a permanent desert.

To geologists, *permanent* is a relative word. Arid climates do not remain monotonously dry for all time. Satellite photography has revealed patterns of river channels of a more humid Sahara that existed not too many thousands of years ago. Climatic change, especially on an ice-capped Earth, is the rule; it is the consequence of an array of interactions among the oceans, atmosphere, land surface, and orbital variations in the Earth's path around the Sun. We know that the Earth was glaciated near the end of the Paleozoic era, and it seems likely that at least small ice caps persisted during the earlier part of the Mesozoic era. Whatever the ultimate causes of climatic change, the desert of the Southwest went through humid and arid periods for the next 100 million years. In a typical cycle, sands deposited during earlier parts of this long time interval would get buried, slightly cemented into sandstone, then be exposed to erosion—producing more sand for the wind to blow. Reptiles and early mammals migrated in and out as the climate dictated. But even this remarkably persistent desert environment did not last.

Invasions of seas flooding larger and larger areas of the continents came closer and closer. As the continent drifted away from

Africa and Europe and moved farther from the equator, the region became more temperate. By the early part of the Cretaceous period, the climax of the dinosaurs was approaching and a wide seaway was spreading over the western interior of North America. By this time, too, high mountains had thrust up along the western borders of Colorado and fed sand, gravel, and mud to rivers and the inland sea. The desert episode was over.

Deserts are neither rare features of the Earth today nor extremely irregular episodes of the past. What makes them unusual is the preservation of the rocks that record past deserts. The sandstones of the Colorado Plateau, like those of the Sahara, are a thin plaster on the surface of a continent. The erosion that sculpts the canyons and gives us the remarkable colors and shapes of spires, slabs, and walls is destroying the evidence. In 10 million years we will have much less of these desert sands to look at. Because the plateau is still in a tectonically active region, future mountains may uplift it to new heights, bringing deeper rocks of different aspect to exposure and erosion. Most if not all of the desert layers will then be gone. The geological forces of an active planet destroy the old as they create the new, preserving less and less of the record as it gets older and older.

Alluvial and eolian sands are deposited over wide swaths of continent and tell us much about the Earth's surface at their time of deposition. The sands at the edges of the continents tell a different story, one of interaction between the border of the continent and the sea that restlessly beats on the shore. In the next chapter we explore how that interaction controls the dynamics of beach sand.

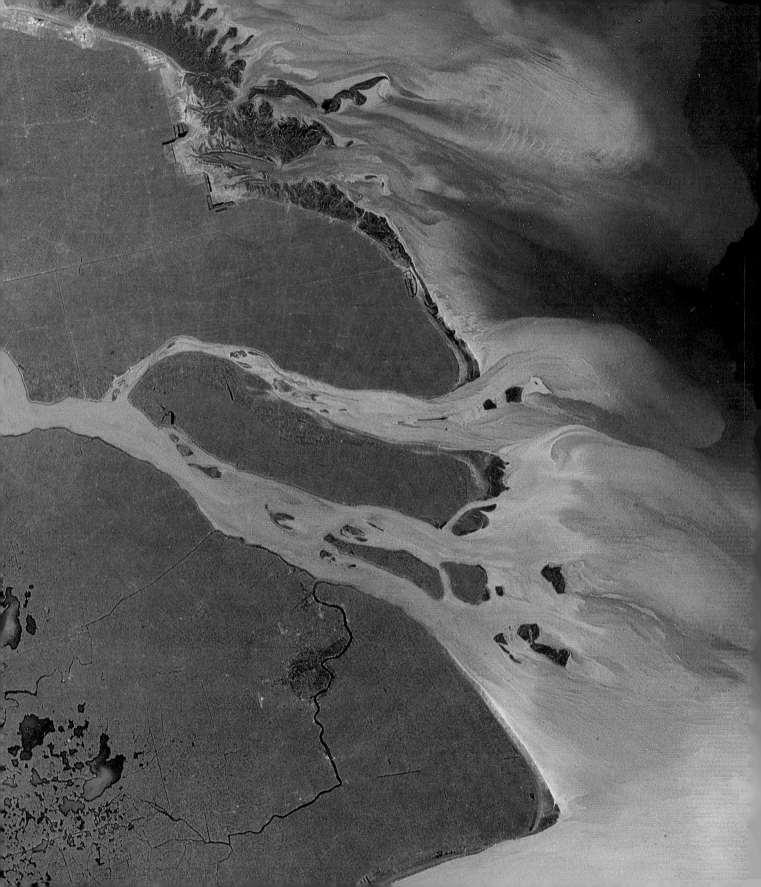

CHAPTER FIVE

Sand at the edge of the sea

The Walrus and the Carpenter
Were walking close at hand
They wept like anything to see
Such quantities of sand:
If this were only cleared away,
They said, 'it would be grand!'

The lament of Lewis Carroll's two beachwalkers is like that of real estate developers: houses built on sand may not last long. While most of the world's population live along the coasts, relatively few live near sandy beaches. Many oceanfront hotels that need beaches to attract vacationers have to truck in sand yearly. What determines the distribution of sand along the shore? Though it may seem random, sandy beach distribution is orderly; it results from the pattern of currents at the edge of the ocean interacting with the supply of sand.

For sand traveling down river-transport systems from its mountain source, the end of the line is the sea. At the sea's edge the waters and winds of the land transfer sand to the waves, tides, and currents of the ocean. The fundamentals of particle transport are unchanged, but different environments have very different flow regimes. Unlike the river's unidirectional flow, the swash of the waves moves the sand up the beach and the backwash takes it out again. The tide flows onto the tidal flat and ebbs to the sea. Currents in deeper waters are more elusive than those of shallow waters, difficult to measure, and less steady. Waves and ocean

Delta of the Yangtze River, China

A part of the western North Atlantic Ocean off the coast of the northeastern United States and the Maritime Provinces of Canada. The Sohm and Hatteras abyssal plains are large flat areas of turbidity current deposition.

currents pile up sand as rippled bands along the shore, wide ribbons extending over the shallow continental shelf, and sheets of muddy sands on abyssal plains, extensive depositional flats of the deep ocean off the edges of the continents.

The currents initially drop most of the sand near the shoreline, then rework and distribute it across the continental shelves. Some travels beyond shallow waters to the continental slopes or the deep basins of tectonically active continental margins, such as those of southern California (Chapter 7 discusses plate tectonics in detail). In a general way, the petering out of sand away from the shoreline reflects the decrease in current velocity and the increase in depth, our two master variables of flow. Responding to these changes is a decrease in grain size, from coarser sand at the beaches to finer sand far out under the deeper waters of the continental shelf.

The view I have just given is that of a satellite observer who cannot see more than the general picture. For us on the ground, the sometimes bewildering detail of shorelines may obscure the ways in which individual environments fit together in a hierarchy. The set of coastal environments includes rocky shorelines, long beaches, salt marshes, tidal flats, and deltaic environments. Each of these environments may be divided into subenvironments. One approach to spelling out the hierarchy is to follow the preliminaries of the broad view I have just given with the analysis of two shoreline environments: river deltas and beaches and bars.

DELTAS

Rivers initially transported to the sea's edge about 90 percent of the sand found anywhere on the bottom of the oceans, and winds carried the rest. The bulk of this river-transported oceanic sand was first deposited at deltas. Geologists working on ancient sandstones know that some of the thickest piles of sand known were laid down at the ancient deltas of major river systems, which persisted a remarkably long time. The Mississippi River delta has continued for 100 million years, slowly building up the land from Cairo, Illinois, to its present mouth south of New Orleans. Its 1000 miles of extension translates to an average rate of a little over 0.5 in. (inch) per year. Curiously, this time scale is about the same as that for one of the basic mechanisms of plate tectonics, the sea-floor spreading that opens up oceans, so that in a few places, sedimentation keeps pace with plate tectonics. Regardless of the rates or the locations, all deltas are united by a common characteristic: how the river flow operates as it reaches the sea.

The simple geometric model that fits all deltas is a steady flow coming out of an orifice into a still basin. The backyard version of this model is a garden hose flowing into a wading pool. If you experiment in your own yard, putting the end of the hose just barely below the surface and pointing straight out, you will discover that the flow continues initially almost unabated into the pool, but it then decays rapidly, spreading out in a widening parabola and entraining surrounding water at the edges. At some distance from the orifice, if the pool is large enough, the flow decays to imperceptible motion.

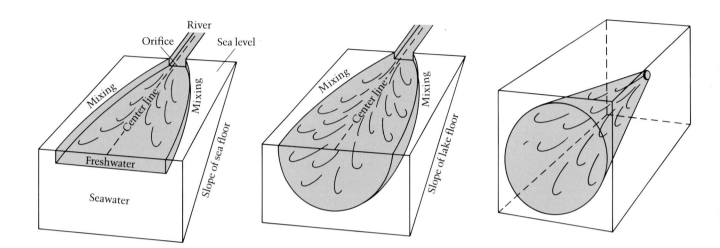

Jet flow describes the gradual spreading out and dissipation of a river current as it enters a lake or sea. *Left:* A plane jet forms when the fresh water of a river (density about 1) flows over seawater (density about 1.02). *Center:* Half of an axial jet forms where a river enters a freshwater lake. *Right:* A full axial jet forms at the central orifice of a swimming pool.

We can describe jet flows like these by three variables: incoming velocity, size of the orifice, and depth of water in the still basin. For example, axial jets take the form of a three-dimensional cone spreading out from an orifice in the middle of one wall of the basin, like the inflow to an olympic-size swimming pool. Planar jets form a two-dimensional fan, where the depth of the basin is little more than the diameter of the orifice and the flow is confined by the boundary, like our wading pool with only a few inches of water in it.

The major dynamic of a jet flow is the mixing of the flow with surrounding still water at the edges of the flow. As momentum is shared with ever-increasing volumes of water entrained by the jet, the velocity slows. The more turbulent the flow, the more intense is the mixing in vortices at the widening edge. The difference between axial and plane jets is in the surface area over which the jet mixes with surrounding water. Because the plane jet exchanges momentum over a much smaller area than the axial jet, its velocity decays more slowly. If the entering flow is carrying sand of a certain range of sizes, it will drop successively finer fractions as the velocity decreases. Because the velocity attenuates faster along the edges than along the center line of the flow, coarser particles are carried farther out along the center line than along the edges. Knowing these facts, we can predict the form of a sand deposit dropped by the jet as a crescent bar, with both its points closer to the orifice. The grain sizes of the bar sand should

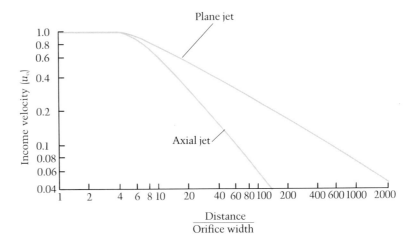

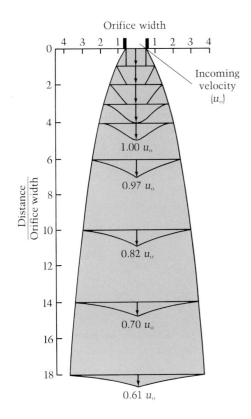

Decrease in velocity of a plane and axial jets.
Above: The distribution of velocities in plane-jet
flow shows the parabolic shape of the jet and
velocities along the center line of the flow. *Right:*
Center-line velocities of a plane and an axial jet,
plotted against the ratio of the downstream
distance from the orifice to the orifice width.

be coarser on the orifice side, and on simple, small deltas, they
are.

We can check this prediction on a larger scale by looking at the
Mississippi Delta, which has been studied more extensively than
any other river mouth. It seems typical of one major kind of delta
and should be a reasonable test for our model. The orifice is slot-
shaped, and the depth of water beyond it is not much deeper than
the bed of the river, so we have a good approximation of a plane
jet. We also have a good tracer of the jet in the muddy color of the
water, which contrasts sharply with the clearer blue waters of the
Gulf of Mexico. In the plume of muddy water we can clearly see
the flow extending far out to sea and the width of the mixing
zone. We confirm our prediction of the bar in the shoals, long
known as a hazard to ship pilots, just offshore of the mouth. Because the Mississippi breaks up into three different channels at
the Head of Passes, we can check these effects at the mouths of all
the passes. All show up. Not surprisingly, the finer-grain silts and
muds are deposited farther out than the sands, tens of kilometers
out to where the jet has completely decayed. Beyond the delta's
apron of sand, silt, and clay, the finest clay is carried far into the
Gulf of Mexico's open waters, slowly settling out to form the
deep-sea clays of the central Gulf.

The Mississippi River delta gives us a way to test the model and
increase our understanding of how deltas work. Yet this simple

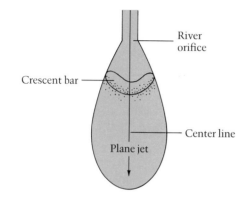

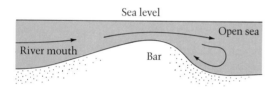

Left: Map and cross section of a crescent bar off the mouth of a river delta, formed as a result of the decrease in jet-flow velocity. *Right:* Satellite view of the lower Mississippi River delta on infrared sensitive film, on which vegetation appears red, relatively clear water appears dark blue, and suspended sediment appears light blue.

model is incomplete. The Mississippi and other deltas are huge tracts of sediment lying landward, seaward, and to the sides of the actual river mouth. At such large scales of analysis, we need geological models that incorporate long time scales as well as large regions. On the basis of past performance, can we predict behavior through time, as the Mississippi continues to grow outward and create new land at the border of the Gulf of Mexico?

We can start to answer that question by following the shifting depositional surface through time. At any instant—to a geologist an instant may mean this year—the depositional surface of the delta's submarine area slopes from the channel and associated floodplain deposits along the center line of the jet to the crescent bar. From there it follows a more gentle slope of a few degrees. Seaward of the bar, the bottom is mantled with finer and finer sediments the farther out we go. Assume that at the next instant the river deposits another thin layer with no change in any parameter of flow. The depositional surface then will move outward, shifted horizontally at the same level, which is controlled

Longitudinal and transverse cross sections of the Mississippi River delta distributaries. As the river sediment is deposited as sand bars, silts, and clays, the delta builds out into the sea, producing bar-finger sands by successive increments added to the cresent bars.

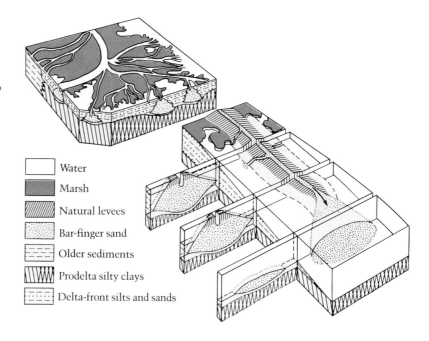

Water

Marsh

Natural levees

Bar-finger sand

Older sediments

Prodelta silty clays

Delta-front silts and sands

by sea level, by a small distance. Much of the total annual particulate sediment load of the Mississippi, about 470 million tons, is distributed along this profile, with smaller amounts of fine clay carried out to the deep sea.

If we now continue this process for a long time we will build up a geological section along whose horizontal dimension at all points we see a constant vertical sequence of sediment types as the locus moves seaward. The farther seaward we follow the bar sand, for example, the younger it is. Above a base of fine clay that was deposited in the deeper waters of the open Gulf lie progressively coarser muds and silts that were deposited closer to shore, then the sands of the crescent bar, then the coarser sands of the river channel. The low-lying land of the deltaic plain is built up by river floods, channel migration, and occasional flooding by tidal surges during storms and hurricanes. Between the multiple distributary channels are salt marshes, where aquatic vegetation traps fine sediment, gradually building the marshes into dry land.

Sedimentologists hunting for facies that will be a key to environment recognize the coarsening-upward sequence and associ-

Top: A continuous seismic profile, or cross section, off the shore of northwestern Africa. The reddish brown lines are seismic wave reflections from various sediment layers; green lines are faults. This profile records the deposition of successive groups of sediments forming the continental shelf. *Bottom:* Cross section of the pattern of coastal sedimentary sequences, as revealed by seismic profiles, that is produced by changes in sea level. The boundary between land and sea deposits, the shoreline moved by progradation from position 1 to position 2 as sea level rose, then dropped to position 3 as sea level fell, and went to position 4 as sea level rose to its present level.

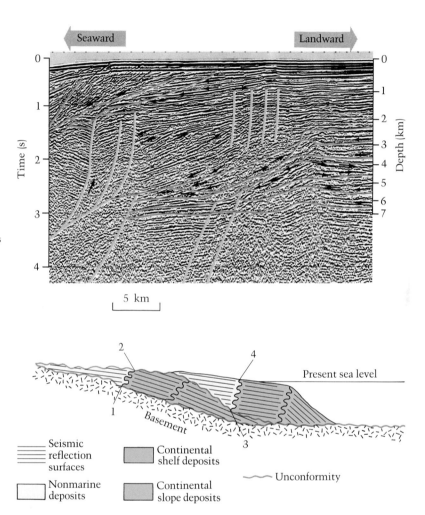

ated land and marine environments as the product of the delta's progradation, or forward motion into the sea, especially if they find marine fossils in the finer-grain layers of the sequence. The coarsening-upward sequence contrasts with the fining-upward cycle of river deposits. The deltaic cycle is formed by progradation into the sea, whereas the alluvial cycle is laid down as the river meanders at any given place. Complications arise that

A salt marsh at low tide, showing the meandering pattern of tidal channels and their drainage network. Cape Cod, Massachusetts.

require additions to the deltaic model. Sea level rises and falls in the many millions of years of a delta's history. It rises with respect to the delta as the Earth's crust below the delta slowly subsides in response to the load of sediment dumped on it and to other tectonic effects at the continental margin. The delta builds both outward and upward in this common situation. Global sea level may rise and fall during ice-capped periods: it falls as water from the oceans is withdrawn to form expanding glaciers and rises when the glaciers shrink. Global sea level may also change as the average rate of sea-floor spreading at midocean ridges increases and decreases. With high rates of spreading, large volumes of sea-floor crust are swollen from the rapid supply of heat from the interior and raise sea level by displacement. At low average spreading rates, the midocean ridges are cooler and lower, allowing the ocean basins to hold more water and lowering sea level. Our deltaic model can predict all these sedimentologic consequences of sea-level changes.

During sea-level lowering the continental shelf is partially uncovered so that the river extends its channel over earlier deltaic deposits. The difference between this sequence and the normal delta sequence is the abrupt transition we see when sea level lowers suddenly, for example, when continental glaciers grow and advance over a period of a few thousand years. This relatively fast transition may be accompanied by the river trenching its bed in the upper reaches of the delta plain, leading to erosional loss of some of the section.

Sea-level rises flood the deltaic plain and push the mouth of the river upstream, producing an inverted sequence of submarine delta deposits overlying the river-channel and land deltaic-plain deposits. Sedimentologists have learned to recognize these transitions in continuous seismic profiles across deltaic margins. The profiles are constructed using high-energy compressional waves generated by trucks on land and ships at sea that filter and enhance the profile of seismic waves reflected from various layers buried beneath land and sea.

Another aspect of delta building is the change in the geometry of the drainage network. As a river channel approaches the sea, its slope to the sea decreases and the river reverses its treelike branching pattern of tributaries. Over the major part of its course, the river is a collection network as it is joined by tributaries and

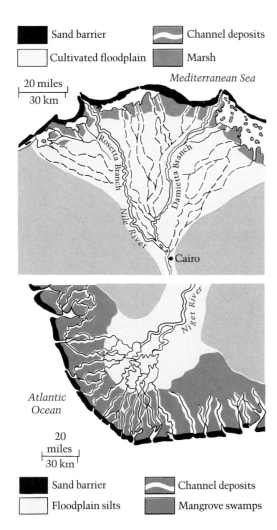

Sand barrier
Cultivated floodplain
Channel deposits
Marsh

20 miles
30 km

Mediterranean Sea

Rosetta Branch
Damietta Branch
Nile River
● Cairo

Niger River

Atlantic Ocean

20 miles
30 km

Sand barrier
Floodplain silts
Channel deposits
Mangrove swamps

Two wave-dominated deltas, in each of which shapes are controlled by waves and longshore currents. *Top:* The Nile delta of northern Africa. *Bottom:* The Niger delta of western Africa.

increases its discharge of water and sediment. When it reverses, it goes into a distributary pattern in which the channel bifurcates, dividing the flow into many smaller channels. The Mississippi below New Orleans divides into three main passes. Other deltas may divide and subdivide into many channels. Before it reverses, the river may break out of its main channel and find a shorter—and therefore higher-gradient—path to the ocean. The Mississippi has been trying to do just this at its junction with the Atchafalaya River well above New Orleans. So far, it has been frustrated by the U.S. Army Corps of Engineers, who want to prevent New Orleans and present shipping lanes from being landlocked by a wayward Mississippi that has adopted the Atchafalaya as its main channel to the sea.

For most of its history, the Mississippi did break out, abandoning a former delta lobe in favor of a new one, to which it would stay stable for only a few thousand years. The whole broad delta region looks like a fan of randomly arranged leaves. As the active lobe is abandoned it continues subsiding as marine muds transported along the shore by currents and tides infill its shallow waters. Here too, the sequence signals the events that produced it: marine clays overlying active delta deposits.

Not all deltas are like the Mississippi, and some rivers, for good reasons, have no deltas at all. Our third condition for jet flow was a still basin. Thus, if the lake or ocean basin has currents, the jet may be modified or dissipated in too short a distance from the orifice to permit a delta to form. This is especially so for smaller rivers whose outflow is weak, compared with the force of disturbing ocean currents. In the natural world, tides or waves produce the currents that modify or prevent delta building. It is no accident that the Mississippi delta formed in the Gulf of Mexico, where the low waves and longshore currents gain energy only during hurricanes and tropical storms and generally have insufficient power to disrupt the jet flow from the river's mouth. Tides in the Gulf are low and tidal currents weak. We see the same regime in the delta of the Rhone River on the Mediterranean coast of France. There the Mediterranean has a relatively short distance of wind travel over the water, and that short fetch minimizes waves; Mediterranean tides are negligible too. The Nile delta, in contrast, is wave-modified.

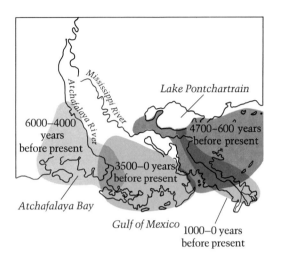

The Mississippi delta lobes of the past 6000 years, a series of major distributing centers that shifted with time.

On both sides of the Atlantic, deltas are missing from large rivers: from the Saint Lawrence, the Hudson, the Gironde of France, the Congo of Africa, and the Amazon of South America. Strong waves and high tides sweep away the sand and mud as it is brought from the river, carrying it along the shore by longshore currents. Downcurrent from these rivers are long, sandy beaches, sand bars, and sand barrier islands. Their deltas have been smeared out as jet flows from these rivers are deflected along the shore. The Mississippi delta also yields some of its sands to longshore currents; even though these waves and currents are too weak to disrupt the jet flow, they have enough energy to create a strong longshore current along the Louisiana and Texas coasts west of the delta. The sands of Matagorda and Padre, barrier islands along the Texas coast, are in part derived from the Mississippi delta and in part from sands of the Sabine and other rivers of southern Texas. Those rivers have no deltas of their own because the longshore currents have carried their sands down the coast.

On coasts where tides and waves are energetic but more moderate than the Atlantic, deltas form with intermediate characteristics. The sand bars may be aligned parallel to tidal currents, frequently at right angles to the general trend of the shore. Recognition of these types of delta is difficult because the complex interplay of river outflow, waves, and tides leaves indistinct vertical or horizontal signatures of sand, silt, and clay beds.

We can recognize some undoubted deltas of the past, however. In North America, the most celebrated is the Catskill delta of eastern New York State. In 1914, an eminent Yale professor of geology, Joseph Barrell, wrote of his idea of a great delta of the Devonian period, about 370 million years ago, and of the high mountain chain shedding detritus that must have lain to the east. It was the sandstones of this delta that have eroded to give the hilly scenery of the Catskill Mountains, which at the time of Barrell's field work had become the preferred summer retreat of the immigrant population of New York City. Barrell called it a delta before we knew about jet flow, the fining-upward alluvial cycle, or the coarsening-upward deltaic cycle. He used only a few criteria, plant and animal fossils and the textures of the sands, to guess its deltaic origin. The sands of the eastern section contained land plants, which confirmed its origin as an alluvial plain. To the west the sands became finer and interbedded with muds containing marine fossils. He guessed that the area was a big pile of sand that thinned to the south and probably to the north, and his guesses were prescient. Since Barrell, the generations of geologists who visited the Catskills as students on field trips and stayed to do further research have mapped the details of the land and marine facies, deciphered the composition of the rocks that were eroded from the eastern mountains, and made estimates of the paleoclimates. The Catskill delta has been placed in the context of the plate-tectonic motions that resulted in the collision of North America with Northwest Africa. That plate convergence, which was the first event in the assembly of continents that formed the supercontinent Pangaea, raised the high mountains that fed the delta with sand, gravel, and mud as it eroded.

In spite of all the successful work on the landward and seaward facies of the Catskill delta, we cannot with certainty identify any beach sands that may have separated continental and marine areas. Sands that were deposited on beaches formed by longshore currents along the delta front have not been mapped with assurance. The reason may not be because they were not deposited there. They may have been deposited far to the north and south and subsequently eroded. In addition, beach deposits are not easy to recognize in ancient rocks. Why? To answer that question we have to evaluate the forces that make a beach and the kinds of sand that are found in what would seem to be one of the most distinctive of environments.

BEACHES AND BARS

Of all environments, the beach seems the most changeable over the short times that humans observe them. The waves beat on the shore with different frequencies as surfers wait for the big ones to come every so many minutes. The tides submerge and expose bars and terraces twice a day. A storm can change the appearance of a beach in a few hours. The beach in winter may be unfamiliar in contour to the summer vacationer. Those who pore over old maps are shocked to find the extent of growth or shrinkage of beachfront in a single century. The Mont-Saint-Michel of France, in medieval times an island at high tide, is now connected with the mainland by a sea of sand brought in by the strong tides.

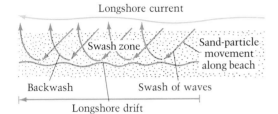

An oblique approach of waves to the shore produces trajectories of sand grains and water that result in a longshore current in shallow water and longshore drift of sand along the swash zone.

Our interpretation of human history and prehistory changes too, as we learn how transitory were the shorelines of antiquity. Troy, overlooking a large arm of the sea in Achilles' time, is now far from the shore. Three thousand years from now historians may have to reconstruct the geography of a disappeared Cape Cod to understand Thoreau's writings about a nineteenth-century ocean beach.

A complete model for beach dynamics requires the interaction of waves, wind, tides, the sand, the physical properties of the rocks or sediments behind the beach, and the shape of the shoreline. The outer beach of Cape Cod is a wonderful place to explore such a model. The waves are moderate to strong, the tidal range moderately high, about 3 m, and the winds variable. It is frequently stormy in winter and quiet in the summer. Medium-to coarse-grain sand forms the beach, and the bluffs or dunes back of the shore are made of easily erodible sands and clays.

Seated just above the swash of the waves, we can directly observe the wave height and frequency in the surf zone. Waves are typically about 1 m high and break every few tens of seconds. The waves breaking on the shore throw sand into suspension and wash it up the beach before the backwash carries it down again. It appears to be at steady state for the short time of an hour at mid-tide, when there appears to be no net buildup or erosion. Looking to the right and left we can see that the waves break traveling along the shore because of the wave crests approaching the shore at an angle. If you watch a piece of shell wash up and down, you can see it move along the shore in a series of parabolas as the asymmetrical swash and backwash transport it up and down, with only a small net movement in the direction of the open angle between shore and wave front. The angle of wave approach therefore determines longshore drift of sand, and the angle itself is determined partly by the direction of the local wind. More critical to the angle is the location of the distant areas from which the waves in the open sea are propagated. Waves are built and sent out in all directions from the strong winds at a storm center, with the waves eventually reaching the shore in the same way ripples do that radiate from a stone thrown in a pond.

The angular approach of the waves to the shore and the return of the backwash water underneath impart a net flow of water along the shore in the same direction as the longshore drift of

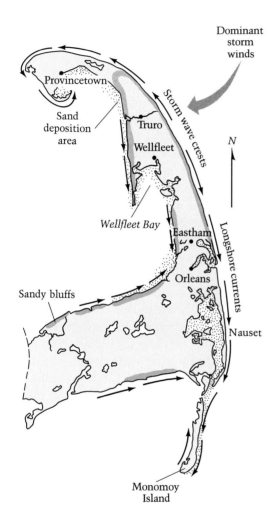

Dominant storm winds

Storm wave crests

N

Longshore currents

Provincetown

Sand deposition area

Truro

Wellfleet

Wellfleet Bay

Eastham

Orleans

Nauset

Sandy bluffs

Monomoy Island

Longshore currents and sand drift on outer Cape Cod work together with erosion of sandy bluffs by storm waves to move sand north and south on the eastern shore and produces bays and spits of sand on the western shore. The Provincetown and Monomay Island areas have been built up in this way.

sand. Swimmers can sometimes feel this longshore current as a tug, but they more frequently take note of it when they end up many hundreds of meters downshore from where they started swimming after only a little effort. A current this strong transports large quantities of sand in shallow waters. Longshore currents and longshore drift often switch directions as the wind shifts or as changing weather fronts determine the offshore disturbances in which the waves originate. On the eastern shore of Cape Cod, which faces the open ocean, the low to moderate waves generated by average storms hit the shore from either the northeast or southeast. The most powerful waves over prolonged storm periods come all from the same direction, the northeast, while the winter northeasters, infamous in New England, howl for days.

If we superimpose this predominant wave pattern on the map of the Cape, we can see what longshore transport leads to. In the vicinity of Truro and Wellfleet, because the bent arm of the outer Cape faces directly into the wave approach and curves away to the northwest and southwest, longshore transport moves sand to the south along the southern part of the beach and to the north along the northern part. We should expect sand to be building up in those directions, and checking old navigational maps shows exactly that pattern. Most of Provincetown is built up of sand transported north along the strand. Nauset and Monomoy Island to the south are the dumping grounds for sand transported to the south. A further check: the high bluffs made up of the original glacially deposited material of the Cape extend only from Nauset to North Truro. North and south of those points we find only low dunes and no glacial material.

If the sand is drifting along the shore and everything seems to be moving out of Wellfleet and Truro, how do their beaches, favorites of Cape Codders, survive? Tragedies reported by the local papers every few years point to the answer: as some children are playing at the foot of a sand cliff 50 m high, suddenly, with a ground-shaking rumble, a collapse of tons of sand buries them. The cliff has been weakened by water infiltrated during the rains of the days before and undercut at its foot by waves at high tide driven by the winds of the storm. Sooner or later it gives way. The sand on the beaches comes from the erosion of the cliffs. We can confirm it by looking at the composition and texture of the sand

Sandbars on Cape Cod produced by longshore currents. The pattern of sand waves shows the dominant current in this view to flow from upper right to lower left.

on the beach, a good match for that in the cliff. Clay in the cliffs is quickly winnowed out and carried away by suspension in the water. Huge rocks dropped by the glacier appear in the cliffs as they are worn back, then drop to the beach, and in a few months or years are covered by the sand and waves. The beach does not get wider this way, for it is in a state of dynamic equilibrium: the longshore transport carries away the sand in amounts approximately balancing the amount supplied by erosion of the cliffs.

The rate of retreat of the cliffs is astonishing. One day some years ago, I suggested to a friend that he had better move his cottage perched at the top of the bluff back a good distance if he didn't want it falling down. With mock precision (I know well the average erosion rate) I estimated the distance to the edge and predicted the event would happen in five years. Five years later exactly, the great storm and blizzard of 1978 did the job. My luck as a prognosticator was just that. No one can predict that specifically. Over more than 100 years of records, in particular the number of relocations back from the shore that ship rescue and coast guard stations had to make during the nineteenth and twentieth centuries, the average rate of erosion is about 1 m per year. Ten thousand years ago, when the glacial age ice was finally melting away from the Cape, the shoreline was something like 10 km east of its present position. Now it lies buried beneath deep offshore waters. In a few thousand years from now, most of the outer Cape will be gone, with a few islands and shoals the only relics of the Pilgrims' first landing place.

A complete model for Cape Cod would show how changing weather patterns, which control the spectrum of wave heights and directions, interact with the regular rise and fall of the tides, from neap tides of less than 3 m to spring tides more than 4 m from low to high. Like many other geological processes, much of the work is done by the moderately intense storms that come only a few times a year, especially those that come during high tides. Hurricanes may do even more in regions where they are strong and frequent, such as the beaches of Florida or the outer banks of the Carolinas. At the other end of the scale, beaches in parts of the Mediterranean with no tides, infrequent storms, and low waves may be relatively stable, with low rates of longshore movement.

Comparison of the Wellfleet vicinity on Cape Cod in 1887 (*right*) with today (*opposite*). Large coastal areas around Wellfleet Bay have been filled in by sand, silt, and mud brought in by longshore and tidal currents in the last 100 years. In the same period the eastern ocean shoreline has retreated about 0.1 km.

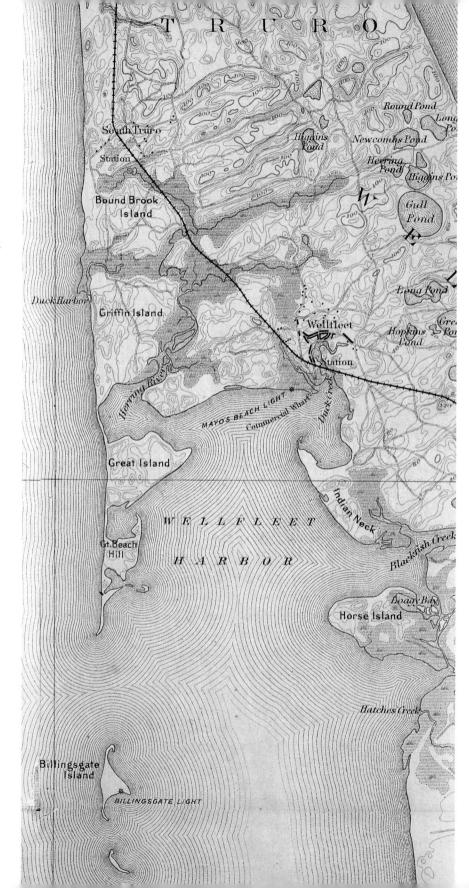

The results of beach erosion, West Hampton Beach, Long Island, New York.

Sand for a beach has to come from somewhere—if not from easily eroded soft cliffs, then from nearby rivers whose sediment load is carried along the shore. Longshore transport from the Mississippi delta and the rivers of the Texas Gulf Coast carry the sand that forms the extensive beaches and bars of Padre Island and other parts of the Texas shore. But elsewhere, where we find little or no steady supply of sand, we see narrow beaches or rocky coasts. For example, the little sand that accumulates in Maine forms only small crescents in protected bays; most of the shore is devoid of sand. In other places the beaches are formed of shingle, pebbles and cobbles too heavy to be carried far.

The complexity of sand transport and deposition on beaches complicates our efforts to identify beaches of the distant past. Studying modern beaches has taught us the pattern of both gently inclined and more steeply angled cross-bedded sand that characterize both the beach proper and the shallow bars exposed at low tide. Yet attempts to prove a distinctive character to sand-grain sizes, shapes, or surface textures have been at best only partly successful. We get the best clues for identifying ancient beach deposits when we can associate different facies with specific shore environments. Shoreward of the beach lies a belt of dunes formed by the winds blowing onshore. Seaward of the beach lie the bars and shallow-water sands that may contain shells of molluscs and other marine organisms. From these patterns we can infer the beach between.

We know far less of the sands of the shallow continental shelf because we can observe them only from a ship's deck through geophysical sensors or, in shallower waters, by divers' observations. They are distributed in ribbons both parallel and at right angles to the shoreline, with the second type influenced by tidal currents; these ribbons may accumulate in cross-bedded submarine dunes tens of meters high. Nearer the shore, these shallow sands interfinger with broad expanses of sands exposed at low tides. The tides make a strong impression on the sandy sediments of these tidal flats adjacent to the shore.

The sandy or muddy expanses of tidal flats, known well to the world's clam diggers, may extend from the shore for many tens or hundreds of meters. In some places, such as the Waddensea off Holland, they extend for tens of kilometers. Systems of riverlike tributaries channel these flowing and ebbing tides over the flats.

The tidal flats of the Waddensea, Holland, are some of the broadest in the world.

Areas between the channels may be populated with vegetation and a diverse group of invertebrate animal species. Where the density of organisms is high and the nutrients abundant, decaying organic matter uses up all the oxygen just below the surface of the sediment and sulfate-reducing bacteria take over, producing hydrogen sulfide. The hydrogen sulfide combines with iron, usually present in coatings of iron-rich minerals, to precipitate iron sulfide, the finely disseminated material that gives such sands their black color. Clams, snails, oysters, marine worms, and many other inhabitants of the tidal flats turn over the sediment as they burrow for food, ignoring the noxious hydrogen sulfide as long as they can maintain occasional contact with the oxygen at the surface. We recognize tidal sands by these fossils and by the distinctive herringbone cross-bedding deposited by the bidirectional tidal current that reverses itself four times every day.

Shorelines are narrow strips of environments. Along most shores the total width of the dune belt, the beach, the tidal flat, the offshore bars, and the shallow-water sands is rarely more than a few kilometers. How, then, do we account for maps geologists

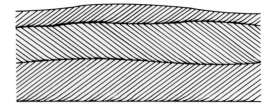

Above: Herringbone cross-bedding produced by reversing tidal currents. *Right:* Cross section of the shoreline facies showing the correlation between sea level changes and transgressions and regressions of the sea over the continental borders.

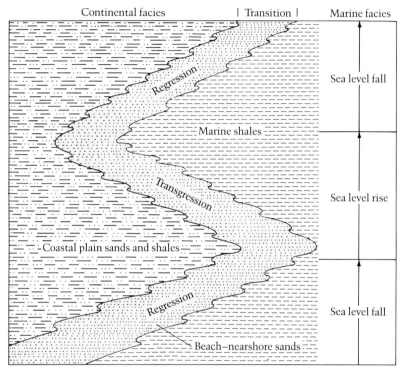

draw that show belts of shoreline facies tens or even hundreds of kilometers wide?

There is no evidence that ancient shorelines were different from today, spreading out over more territory. But over time, the changes in sea level that modify deltas also spread shoreline environments over a wide belt. As sea level rises—owing either to global changes or to local subsidence, a slow sinking of the earth's crust in response to tectonic forces—the shoreline environment moves inland, the sea transgressing over the continent. Transgressions result in an association of sands, from offshore and beach to dune environment, that are facies but not time equivalents. When sea level lowers, the shoreline regresses out to sea, leaving another belt of sands.

When sea-level changes are frequent, so are transgressive–regressive cycles. The waxing and waning of the glaciers of the recent ice ages led to several such cycles in the past few million

Cliff dwellings of the Anasazi, early American Indians who used this canyon of late Cretaceous sandstone for their distinctive architecture from 750 to 1300 A.D. The sandstone was deposited along the shorelines of a great inland sea that periodically transgressed over its coastal plain.

years. Other, longer-term cycles last many millions of years. One of the best-known series of these cycles deposited the sandstones of the later part of the Cretaceous period along the great inland sea that once existed where the Rocky Mountains are now. The sandstone where early Native Americans carved their cliff dwellings in Mesa Verde National Park is a remnant of one of these transgressions. The last of the regressions was permanent, leaving the continent high and dry just at the time when the dinosaurs became extinct—perhaps because the Earth collided with a comet. The cometary impact, according to current theory, caused an enormous explosion that threw so much dust into the atmosphere that the sun was obscured, the sky darkened, and the climate so changed that the dinosaurs could not survive. The disappearance of the seaway, coincident though it may have been with the collision, seems impossible to connect causally. Instead, the flooding and recession of the sea from the continent related to forces operating over a longer time scale closer to home: the plate tectonic movements that produced the mountain chains of the Rockies. The connection between plate tectonics and sand deposition will become much clearer after we explore the sands of the deep sea in Chapter 6.

CHAPTER SIX

Continental margins and turbidity currents

Before modern geological oceanography in the 1930s, sand in the deep sea was a simple subject: there was none. Geologists at that time could explain sand distributed at the edge of the sea and across the relatively shallow continental shelves but thought it obvious that currents strong enough to carry loads of sand could not exist in waters several kilometers deep. The conventional wisdom of a quiet, dark deep persisted long after the *Challenger* expedition of 1872. That first major exploration of the ocean floor had revealed a varied, complex sea floor, with great abyssal plains bordered by extensive hilly regions, in turn broken by high submarine mountain ranges, such as the Mid-Atlantic Ridge. The *Challenger* dredged from the deep-sea floor not only the abundant muds but many samples of sandy "oozes" made of the tiny calcareous shells of one-celled organisms, the foraminifera. In addition, samples of interlayered silts, sands, and muds were picked up from some of the abyssal plains close to the continents. Although the *Challenger* discoveries had an impact, a long sleepy period of ocean study followed, in which few advances were made.

In the decades from the 1930s to the 1960s, new coring tools and geophysical profiling allowed rapid exploration of the ocean floor and destroyed our earlier, simple assumptions. For example, we found sand in fans where submarine canyons opened onto the deep-sea floor at the base of the continental shelf; we recovered silt and sand layers by the hundreds and began to learn where sand could be found in the deep sea. Exploration showed more sand in the deep-sea trenches bordering volcanic island arcs and revealed sand and gravel in high-latitude ocean floors where melting icebergs calved from glaciers dropped their coarse sediment

Underwater photograph of a sandfall in the head of Lucas submarine canyon, Baja California, Mexico. Waves driven by large storms carry coarse shallow-water sands into the nearshore canyon head so rapidly that entire bodies of sand slowly flow downslope.

load. What kinds of currents could operate to distribute sand in all these different environments?

DENSITY CURRENTS AND TURBIDITY CURRENTS

The study of turbidity currents brings together an alpine lake, submarine canyons, and perplexing sedimentary sequences found in the Apennine Mountains of Italy. The story starts with F. A. Forel, a Swiss studying the dynamics of Lake Leman at Geneva in 1885. Watching from the bridge over the Rhone River where it entered the lake, Forel was fascinated by Rhone water, which ran turbid with suspended sand, silt, and clay down the sloping bottom of the lake. He could see the steady flow clearly fanning out across the bottom through the limpid surface waters of the lake. (Today we would call it a plane jet, which is described in Chapter 5.) Density currents act like that, Forel knew. If a fluid of a given density is introduced beneath another fluid of lower density along a slope, the heavier fluid will run downhill along the bottom.

But what made the river's inflow so dense that it flowed beneath the lake water? The first answer seemed obvious: the icy cold Rhone waters had drained the meltwaters of alpine glaciers and were therefore denser than the warmer surface waters of the lake. Unfortunately for this attractive idea, temperature measurements in winter and summer showed that the slight difference in temperature between the warmer lake and colder river waters could not be causing the density difference. Forel pondered the problem and then saw an answer: the river water was denser because it carried suspended mud; the suspended particles were so fine and continuously distributed through the water that they increased the density just as effectively as a dissolved substance would. Thus, a new kind of density current was discovered, much later to be termed a turbidity current. Confirmation of these ideas came in the early twentieth century as hydrologists mapped turbidity flows in muddy rivers entering large reservoirs. By the mid-1930s, engineers had worked out the behavior of turbid Colorado River water as it flowed the entire distance down the submerged river channel beneath the quiet waters of Lake Mead, the artificial reservoir impounded behind Hoover Dam.

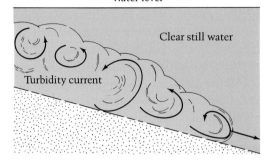

Drawing from a photograph of an experimental turbidity current generated in a laboratory flume by Phillip Kuenen of the Netherlands in 1936.

Forel's work on turbidity currents caused no great stir and remained only a minor curiosity until it was resurrected 50 years later. Then it was proposed as a way to settle a new controversy that had arisen over the origins of submarine canyons discovered along the continental slopes of the Americas and Europe. R. A. Daly of Harvard was a master of theory, experiment, and field observation in dozens of different areas of geology. A bookish scholar too, he had looked at the work on the Swiss lakes, worried about inadequate explanations of submarine canyons, and calculated densities and velocities of possible flows under the ocean. He argued strongly for density currents as the mechanism of erosion of the Hudson and other canyons. Daly, with physics colleague P. W. Bridgman a founder of the Committee on Experimental Geology and Geophysics at Harvard, dismissed offhandedly the possibility of experiment on such a large-scale natural process.

Philip Kuenen of Holland, at that time a young marine geologist who had spent several years with the Dutch East Indies oceanographic expedition, immediately proceeded to show that valid experiments could be done with fairly simple equipment. First with a small laboratory tank, then with a large room whose floor could be flooded, and finally with a long outdoor trench, Kuenen showed that turbid suspensions do flow down slopes underwater, will continue for long distances without mixing with the overlying water, and could erode a sediment bed. Kuenen calculated from his measurements that the head of the turbidity current might have flow velocities between 0.4 and 3 m/s, ample to carry sand grains.

GRADED BEDS AND MIXED FOSSILS

The next strand of the turbidity-current idea was its connection to ancient rocks. For over a century geologists had been familiar with thick sections of marine sandstones and shales that resisted simple interpretation. The sandstones were graded upward, from pebbly or coarse sands at the base through finer-grain sand, silt, and clay at the top. The sands were cross-bedded and rippled, indicating strong current flow. The graded sequence was terminated by a sharp, abrupt contact with the coarse sandstone of the next cycle above. These cycles were repeated hundreds of thou-

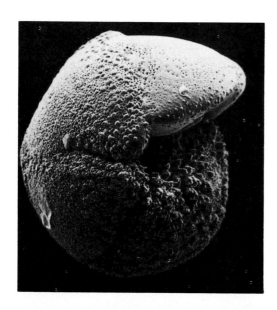

Scanning electron micrographs of foraminifera, tiny marine unicellular organisms that secrete shells of calcium carbonate. These shells may accumulate on the sea floor as forminiferal oozes.

sands of times in some mountainous regions, most spectacularly in the northern Apennine Mountains of Italy. The rocks contained marine fossils, many of them floating forms characteristic of the open sea. The cycles were clearly different in origin from the alluvial fining-upward cycle deposited on land (see Chapter 4). An Italian geologist, C. I. Migliorini, knew these rocks had to be the product of deep undersea currents; in a superb collaboration, he worked in the field with Kuenen, comparing modern and ancient sediments to see the many similarities.

Sandstones like these posed a different problem for oil company paleontologists near Ventura, California, who were studying shells of foraminifera, microscopic unicellular organisms, to establish the time sequence of the rocks to aid in oil exploration. The shells told a baffling story. Some of the foraminifera were benthonic, that is, living on the sea floor, with different species characteristic of different depths of water. The assemblages of shells in these sands, however, contained foraminifera of different depth zones mixed heterogeneously. In some of the thin shales at the tops of the graded sequences, they could find only pelagic foraminifera, those living in surface waters. What kinds of current could transport both coarse sand and fine mud in a graded sequence, mix up benthonic foraminifera of different depths and floating forms, and turn off and on thousands of times? Geologists, paleontologists, and oceanographers argued and tried to use storms as a mechanism, but no one could see how storms could affect waters hundreds of meters or even kilometers deep. The central fact was that deep-water sand with abundant cross-bedding and ripples evidenced strong currents.

The whole picture came together about 1950 with the correlation of the results of Kuenen's continuing experiments, his and his students' field work on graded beds in the Alps and Apennines, ample evidence from foraminiferal studies, and new studies of bottom sediments of the oceans. Cores of sediments on flat abyssal plains showed the same association of sediment types: sandy beds, including mixed foraminifera, alternating with fine muds in graded beds.

The cap of these developments came with the retrospective analysis of a series of major transatlantic cable breaks following an earthquake on November 18, 1929, off the Grand Banks of Newfoundland: In the 1950s, Maurice Ewing, the director of

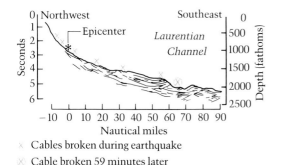

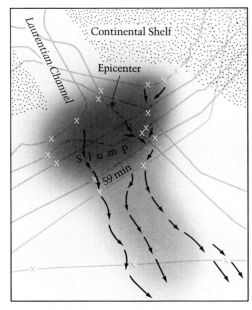

Cables broken initially by slump
Cables broken initially by turbidity current
× Cable break
➤ Turbidity current

Seismic reflection profile and map of the Grand Banks, Newfoundland, area where cable breaks signaled the beginning of a slump and turbidity current.

Columbia's Lamont Geological Observatory, and his colleague Bruce Heezen took time out from their pioneering studies of mid-ocean ridges and rifts to compare the precisely known times of breakage of the cables in dozens of locations on and at the foot of the continental slope. The snaps of the cables on the slope and the abyssal plain below—no breaks occurred on the continental shelf above—took place at times and positions that could be explained only by the passage downslope of a fluid wave triggered by the large earthquake. The quake generated a large slump of clay and sand at the edge of the shelf and the slump, cascading down the slope, formed a turbidity current that raced down at velocities of about 20 m/s. At the foot of the slope the flow fanned out over an abyssal plain. Coring the sediments in the cables' paths, Ewing and Heezen recovered the sandy graded beds that were sedimented by the 1929 cable-breaking event and earlier turbidity currents. In the next few years the Lamont geologists mapped similar cable-break sequences—common events rather than isolated occurrences—on continental slopes off the coasts of North Africa and northern South America. The concept of turbidity currents in the modern oceans was secure.

HOW TURBIDITY CURRENTS WORK

Out of so many apparently disparate observations came a coherent fluid-dynamic and geologic theory. The first part of the theory is the generation of a flow. Steady flows of muddy rivers into freshwater lakes are simple. All you need is sufficient turbidity of the water to make it flow underneath as a density current. The minimum effective density—the difference between the flow density and that of the overlying water—need be only 0.005, and many streams carry enough suspended mud to reach that density.

Muddy rivers entering the ocean do not act that way, however. With the density of seawater at about 1.02, a freshwater river needs a heavy load of suspended mud to exceed it, and few rivers carry that much mud at any time. Those that do face another hurdle, mixing outflow water with seawater. As we have seen, in jet flows at deltas the current dissipates by mixing at the boundaries of the jet. That mixing also dilutes the turbidity and so reduces the density, making it less likely to flow underneath the main body of water. If the waters are still turbid when they reach

Steady turbidity currents start where a turbid river, such as the Colorado, enters a freshwater lake, like Lake Mead.

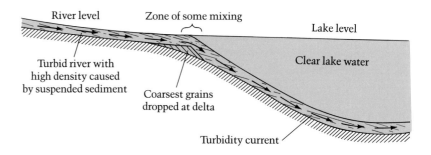

River level

Zone of some mixing

Lake level

Clear lake water

Turbid river with high density caused by suspended sediment

Coarsest grains dropped at delta

Turbidity current

the open waters of the continental shelves, the waves and tides will quickly mix and homogenize the turbid waters with the surrounding clear seawater and dissipate any underflow. The net result is the absence of any turbidity currents on continental shelves under all but the most special conditions. This checks with the absence of graded sandstones in known shallow-water deposits.

A different mechanism for turbidity-current formation operates on the continental slope. Here we can make an analogy to the buildup and cascading of sand down the slip face of a sand dune, as discussed in Chapter 3. At continental margins, the sand and mud transported by currents of the continental shelf settle out as they reach the somewhat deeper waters of the upper parts of the continental slope. There continued sedimentation builds the slope to an unstably steepened angle. Then, triggered by earthquakes, the slope gives way in a slump—a landslide under the sea. The slump mixes sediment and water along the bottom to a heterogeneous, confused mass of flowing sediment, creating a turbid mass of water that flows down as a density current. Although mixing and dilution of the turbid flow with the overlying clear waters might weaken the turbidity current by lowering its effective density, mixing is inhibited by the as yet low velocity of the newly born turbidity current.

As the flow accelerates down the inclined plane of the continental slope, it erodes its base and entrains, or brings along, an increasing sediment load. Just as it does on land, the erosion quickly transforms a thin sheet flow to a channeled flow. Rills grow to gullies and shallow channels to full-fledged submarine canyons. As it erodes along its channel, the flow mixes material

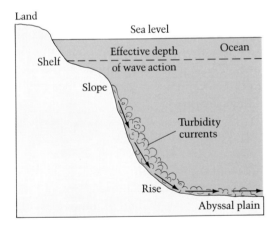

Top: Sandfall at the head of a submarine canyon. These falls generate sandy flows, like turbidity currents, that lay down fans of sandy sediment at the foot of the continental slope. *Bottom:* Formation of turbidity currents at continental margins. Slumps triggered by earthquakes continue downslope as turbidity currents, depositing submarine fans at the openings of the canyon onto the sea floor and abyssal plains beyond the fans.

from different depths along the slope, accounting for the mixed foraminiferal assemblages that had perplexed the paleontologists.

The accelerating turbidity current carries the seeds of its own destruction. Although it mixes with clear water less at the start, the higher the velocity of the flow, the more mixing with clear water at its upper boundary. Waves may form at the interface between turbid current below and clear water above; when those waves break, the flow mixes with clear water more extensively. The mixing dilutes the turbidity and thickens the flow, providing a decelerating resistance that works against the velocity increase expected from the slope alone. When the flow hits the foot of the slope, it shifts out of the balance between the acceleration from gravity and the deceleration from resistance. No longer confined to a channel and facing the flat ocean floor—an abyssal plain created by earlier flows—it begins to decelerate more rapidly.

As the flow loses velocity, it begins to drop its sediment load, first the coarsest material: pebbles, pieces of still-soft sediment torn from underlying beds, and the largest sand grains. Because the flow is highly charged with mud, some of the finest material is deposited too, entrained in the rapid settling of coarse particles near the bed. As the flow wanes and dissipates, successively finer-grained deposits are laid down, giving rise to the graded bed sequences that geologists had noted for more than a century. The settling and entrapment of mud from the dense turbid flow during the whole graded sequence partly explained another peculiarity: the clay matrix between sand grains in the formerly perplexing rocks we now call turbidites. Settling of clay with sand as the turbidity current waned accounted for only part of the matrix; no matter how one stretched the numbers, only a very-low-percentage clay matrix could be produced. The much more abundant clay matrix of some ancient turbidites is owed to a mechanism we will cover in Chapter 8, on postdepositional changes in sandstones.

Just as a river deposits an alluvial fan as it emerges from a mountain front onto a broad valley floor, the turbidity current lays down a submarine fan at the mouth of a submarine canyon at the foot of a continental slope. The fluid-flow conditions are similar: the flow accomodates to the change from a confined channel to an open floor by decreasing velocity. The submarine fan has its own channels. In the same way that a river changes its channel

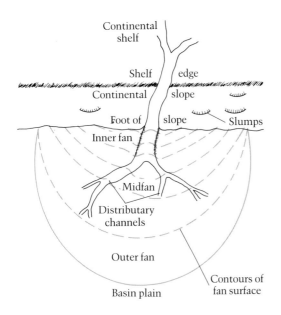

Continental
shelf

Shelf   edge

Continental   slope

Foot of   slope   Slumps

Inner fan

Midfan

Distributary
channels

Outer fan

Contours of
fan surface

Basin plain

Submarine fans form at the foot of the
continental slope, where submarine canyons
open and the turbidity current expands and
loses velocity.

pattern on its delta from tributaries to distributaries, submarine-fan channels of turbidity currents diverge in a branching pattern. As the current decreases down the submarine fan, successively finer sand grains are deposited with a gradually higher amount of silt and clay. At the outer edge of the fan, the current debouches on the broad flat abyssal plain and then spreads out as a sheet flow. As it steadily wanes, the fine-grain sand forms thin layers; more of the deposit becomes silt and clay. At the outer reaches of the plain, the flow lays down only silt and mud.

Sedimentologists have been able to piece together the structure of ancient submarine fans by analyzing the form, texture, and sedimentary structures of the channel and interchannel deposits. Looking at the data, they could make out the indistinct outlines of the channels and overbank deposits formed in the same way as river-flood deposits. Far beneath the waves, the infrequent largest flows overflowed the banks of their fan channels and spread out over the broad surface of the fan. No one has been able to catch such a rare ocean-bottom flow in the act; so far, we cannot install sensors on the sea floor that would last long enough to record such a flow. We therefore know these submarine floods only by the sediment they lay down. These mute sands speak of the turbid flows that made them in a vocabulary of grain size, sedimentary structures, and bedding sequences.

THE TURBIDITE REVOLUTION

The miniscientific revolution of turbidity currents and turbidites was accomplished in an explosion of symposia, friendly and acrimonious debate, renewed experimentation, and much field work in the late 1940s and early 1950s. Classic sections of graded beds, well-exposed sequences that had been described for a century, were reinterpreted in the light of turbidity current theory; hitherto unexplained sedimentary structures, textures, and bedding sequence variations were encompassed by the new theory. A new interpretation of marine sediments at continental margins and abyssal plains, based on turbidity currents, would also play an important role a decade later in the major revolution of plate tectonics. Some of the last recalcitrants were still murmuring 10 years later, "But nobody has ever *seen* a turbidity current in the ocean." That was and remains true. Not that people haven't tried.

Flute casts, one type of sole-marking, on the underside of the basal sandstone of a turbidite. These casts are the record of spoon-shaped scours on the mud over which the turbidity current flowed.

Jacques-Yves Cousteau tried using his submersible to knock off some material from a canyon wall in the Mediterranean, only to raise a small muddy cloud. The late Francis Shepard of Scripps Institution of Oceanography, the dean of American submarine geologists, tried dynamiting in the La Jolla submarine canyon not far offshore from Scripps pier, also without success. Apparently, one has to be in the right place at the right time to observe this theoretical construct at work; it is likely that the evidence will remain circumstantial.

Part of the excitement of field work on turbidites was the translation of turbidity-current mechanics into actual sedimentary structures such as sole-marks, enigmatic forms found at the base of the sandstones. Sole-marks—so called because they were seen as raised ridges and as various bulbous forms on the undersurfaces of sandstone beds—are in reality casts of depressed forms originally on the surface of the shales underlying the sandstone. Viewing them from this vantage point—and you could recognize a sedimentologist looking for them by his neck cricked from trying to look up at the surface—we recognized the various raised forms as casts of grooves, ripples, depressions scoured by currents, and other indentations. Grooves and other structures were aligned parallel to the direction of flow, as indicated by the cross-bedding and ripplemark directions of the sandstone. They were formed as the turbidity current flowed over a mud bed, the last of the previous flow's deposits, sometimes making grooves as it dragged pieces of shells or other large objects along the bottom.

Seeing the significance of sole-marks gave paleocurrent mapping a new impetus. Whereas cross-bedding in alluvial sandstones gave the paleoslope of a land portion of a continent, sole-marking paleocurrents gave the slope of deeply submerged parts of the continental margins. Some of these currents flowed down the continental slope. Others flowed down the walls of deep-sea trenches, then turned to flow longitudinally along the trench. We began to map the topography of ancient sea floors.

At the same time, marine geologists were mapping turbidite deposits on the modern sea floor. Now we understood the significance of the continental rises, aprons of sediment deposited at the foot of the continental slopes that merged at their lower end with the abyssal plains spreading over thousands of square kilometers of the deep-sea basins. Those plains were the surface of the oceanic lithosphere that had formed at a midocean ridge, and they blanketed a hilly topography of basalt. The thickness of the plain sediments was proportional to their age, assuming a more or less constant rate of accumulation. Later, this measure was to be an important confirmation of sea-floor spreading. The older parts of oceanic crust that had traveled farther from the ridge where they were formed were blanketed with a thicker layer of turbidites than the more recently formed parts closer to the ridge.

As we dated turbidite layers and counted the number of layers, we began to get some sense of the rate of accumulation. The spasmodic turbidity currents seemed to be triggered once every 100 years. Because different turbidity flows would come from different places along the edge of the continent, a much longer time interval was needed, where sediment supply was low, for sediment to build up to the point where an earthquake could trigger a slump. We could distinguish the rapid buildups and frequent

Complex sole-markings, including compound flute casts and grooves, on the underside of an early Paleozoic turbidite sandstone that has been structurally deformed to make the bedding vertical.

earthquakes of subduction-zone trenches near continents, where there was a high supply of sediment and tectonic activity, from the slower pace of turbidite sedimentation in trenches far out in the ocean, with the limited land areas of small volcanic islands such as those in the South Pacific. With most of the loose ends tied up, deep-sea research and speculation have now turned to other kinds of currents.

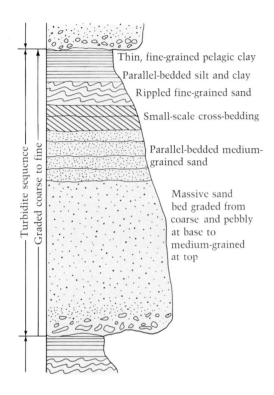

Thin, fine-grained pelagic clay

Parallel-bedded silt and clay

Rippled fine-grained sand

Small-scale cross-bedding

Parallel-bedded medium-grained sand

Massive sand bed graded from coarse and pebbly at base to medium-grained at top

The vertical sequence of a turbidite, the deposit of a turbidity current. Sequences range in thickness from a few centimeters to 3 or 4 meters. Not all subunits are present in every sequence, the exact pattern varying from fan to abyssal plain.

CONTOUR CURRENTS AND BENTHIC STORMS

In 1964 Charles Hollister, a graduate student in marine geology at Lamont Geological Observatory, working with his mentor, Bruce Heezen, reported on sands on the continental slope that were cross-bedded from the currents that deposited them. Contourites, as they called them, were not turbidity-current sediments but sands that were laid down by more normal currents running along contours parallel to the continental slope. Hollister and Heezen had been intrigued by recent observations that confirmed a theoretical prediction by physical oceanographer Henry Stommel at Woods Hole Oceanographic Institution: under the western part of the North Atlantic Ocean ran a countercurrent, a strong flow at depth that ran north to south, the opposite direction to the Gulf Stream at the surface.

The two marine geologists wondered whether the countercurrent might have left a sedimentary trail on the bottom. After searching existing cores and cruising beneath the countercurrent to get new cores, they were able to display the sand cores that proved the point: contour currents ran along the contours of the continental slope, picking up sand and depositing it like any other current. How could such a current be inferred from ancient rocks? The answer came in comparing the directions of sediment transport indicated by paleocurrents and sedimentary structures of turbidites and contourites. In several areas, geologists were able to show that where the turbidity currents ran down the slope, the cross-bedded contourites, which did not show the typical sequences of turbidites, ran at right angles to the turbidites.

For a long time now at Woods Hole Oceanographic Institution, Hollister has pursued the interrelations of physical oceanography and sedimentation at continental margins and in the deep sea. He thinks there are benthic storms, analogous to the storms that accompany the passage of weather fronts on land, that stir up the bottom and transport sandy and muddy sediments. These storms may cause large pile-ups of sediment in some places in the ocean that seem inexplicable otherwise. Land geologists have yet to recognize many deposits, other than a few reports of contourites, that might have come from ancient currents like the ones Hollister has mapped in the modern oceans. The identification is likely to remain difficult, for deep-sea deposits are likely to be preserved on the continents in deformed mountain belts, where sands and shales may be metamorphosed and structurally distorted, not easy places to observe subtle differences in depositional style.

CHAPTER SEVEN

Reading plate tectonics from sand

I t took about 150 years from the publication in 1795 of James Hutton's *Theory of the Earth,* a work that marked the birth of modern geology, before sedimentologists could formulate the theory dominant today: the tectonic control of sandstone compositions. Bits and pieces of the concept had been accumulating from the time of Hutton—a few fragments had even come from centuries before—but only the faint outlines of a theory were discernible in the early twentieth century. Alpine geologists during the last half of the nineteenth century saw what they thought was an important, relatively simple relationship between sandstone types and the faults and folds of the Alps. The dark gray, thinly bedded sandstones interleaved with dark gray shale layers in hundreds of couplets clearly seemed to have been deformed in the building of the Alps. Folded, faulted, and metamorphosed by high temperatures, great thicknesses of these rocks, the flysch, must have been deposited in a deep sea before the main event of alpine mountain building. In sharp contrast to the flysch was the molasse, another series of thick light-colored, coarse-grain sandstones and conglomerates, only slightly deformed, if at all, by the paroxysm that had so affected the flysch.

Because Europeans—and Americans in their turn—tended to make universals of their own particulars, the alpine flysch and molasse were proclaimed by Swiss, French, and German alpine geologists as a general pattern that could be seen in mountains of other ages and different places. In the early stages of a mountain chain's development, a "foredeep" formed in which deep-water sands and shales were deposited, the detritus from source lands within the belt of growing mountains. As this flysch was itself caught up in the crescendoing, large-scale tearing and crumpling

Windblown sand advancing over a sea cliff.

Photomicrographs of two of the commonest igneous rocks exposed to weathering in uplifted mountains. *Left:* A granite. The largest masses are potassium feldspar; smaller clear grains are quartz. *Right:* A basalt. The bulk of the rock, a finely crystalline dark matrix, is recrystallized volcanic glass; larger crystals are plagioclase feldspar.

of the crust, cannibalism began, in which previously deposited sediment was deformed and uplifted, exposed to erosion, and thus recycled to a new generation of sediment, all within the mountain belt. As the mountains evolved they fed on the parts that had formed earlier. Erosion of the high mountains produced a glut of sand, gravel, and boulders, including pieces of the flysch, which had been thrust to the tops of the highest peaks. Mountain rivers brought their freshly eroded load to the lowlands in front of the ranges and deposited the molasse along flood plains and alluvial fans, just as the Rhine River carries the sediment from the Alps today.

In so recent a mountain belt as the Alps, where the general pattern could still be seen, geologists could make sense of the flysch and molasse. In older mountains, such as the Appalachians of North America, long-continued erosion seemed to have destroyed much of the evidence. In contrast to the Alps, with a mere 30-million-year history of deformation, the Appalachians were 400 million years old. Certainly it would be unbelievable to trace such a history from the eroded roots of mountains of the Precambrian, uplifted more than a billion years ago. Yet that is how far back a young geologist in his twenties would go to determine the origin of sandstones and conglomerates shed from a long-vanished mountain belt.

On a canoeing trip through the wild lakes and forests of northern Minnesota and western Ontario, Francis Pettijohn, a graduate student at the University of Minnesota, saw country north of Lake Superior that in 1927 was part of an older time: no tourists,

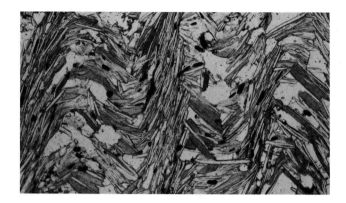

Photomicrographs of two of the commonest metamorphic rocks of deformed mountain belts. *Left:* A mica-quartz schist, metamorphosed from a shale, showing microfolding of the thin flakes of mica. *Right:* A quartzite, metamorphosed from a sandstone; all of the larger grains are quatrz; the finer matrix includes micas and iron oxides.

no economic development, just hunters and trappers. Amid the beautiful lakes, the dense woods, and the never-ending mosquitoes, Pettijohn could see bare, glacially shaped and smoothed rocks of the oldest era of Earth's known history. These Archean rocks, older than 2.5 billion years, showed all the diversity of minerals and textures of the much younger rocks familiar to most geologists.

A particularly striking rock caught Pettijohn's attention on the shore of Abram Lake, only a short canoe ride from Sioux Lookout, an outpost of the Canadian National Railway. Here were conglomerates with cobbles and boulders in the midst of a terrane dominated by enigmatic igneous and metamorphic rocks. To Pettijohn, the conglomerates flashed a sign of rapidly flowing rivers carrying the gravels and sands of a mountain range built in the earliest days of the Earth's history. He decided on the spot that these rocks would be his doctoral thesis—they became even more, occupying him for the next 20 years. Pettijohn returned repeatedly from his teaching post at the University of Chicago to the Precambrian rocks of northern Wisconsin and the Lake Superior country, gestating his ideas on the formation of sedimentary rocks in this early stage of Earth history.

Familiar with the literature of the Alps and up-to-date on the continuing ferment of ideas on the origin of mountain belts, Pettijohn pondered the relationship between rock compositions and tectonics. Tectonics set the stage for sandstone deposition by providing mountainous source areas where wind, water, and ice could bite into ancient sediments—some of them metamorphosed into slates, schists, and quartzites by the heat and pressure of deep burial in the Earth's crust—and granites, basalts, volcanic ash deposits, and other igneous rocks. In an actively eroding mountainous source area, the mineral composition would reflect the speed of mechanical weathering that would allow the feldspar and other unstable grains to be preserved (the decay of feldspar is discussed in Chapter 2). The debris of the mountain tops would be carried out of the range and dumped as molasse. Thus, unstable mineral compositions could be linked to stream deposition during and following the intense stages of orogeny, or mountain building.

The thick flysch deposits Pettijohn had seen in other parts of the Lake Superior country were dark sandstones, altogether different from the molasse. They were like the marine graywackes, also composed of unstable rock and mineral fragments, of the Harz Mountains in Germany. These "dirty" sandstones—dark as opposed to the more familiar light and "clean" quartz sands—were dominated by volcanic rock fragments, feldspars, and a dark clayey matrix, or paste, holding the angular grains together. Graywackes were signs of terranes related to volcanism and perhaps to earlier stages of orogeny. Both flysch and molasse were strongly tectonically determined, in opposition to the quartz-rich sands of stable, low-lying continental interiors.

Pettijohn was not alone: his good friend Paul Krynine had also theorized about the relationship between tectonics and sedimentation, but not with the oldest rocks. Trained at Yale, Krynine wrote his Ph.D. thesis on red sandstones of the Triassic period that were beautifully exposed along the Connecticut River valley north of New Haven. These sandstones had been deposited in a series of rift valleys, long depressions bounded by faults, formed where the crust had been pulled apart. He also had experience with the most recent tectonically controlled sandstones, the Siwalik Series of the foothills of the Himalayas, the highest and youngest of the world's great mountain belts.

Krynine and Pettijohn were a study in contrasts and appeared the more different, I realized, as I came to know them better. Pettijohn, my teacher, was usually quiet and low-key but would quickly prod his students and colleagues to explain their hunches or feelings about the rocks they were observing. Having grown up in small and medium-size towns of the American Midwest, he embodied the then new American tradition that peopled the sciences from the many state universities and liberal arts colleges spread over the states from Ohio to Nebraska. Always kind, he could be a quiet tease, especially at an outcrop. Krynine was a different sort, a young White Russian emigre in whom the European tradition was filtered through the glasses of czarist academics and intellectuals. I first met Krynine when he rose slowly but aggressively to ask me a pointed question when I presented my first paper at a national meeting. In succeeding years, after getting to know him, I saw how he covered a subtle and witty intellect with the behavior of an enfant terrible, an ebullient agent-provocateur. With a heavy Russian accent he layed about him with barbs, proposed dramatic new ideas, and stormed around, stimulating his students and anyone else who would listen to think in new ways. Once, from the floor at a major symposium, he satirized the speakers and gave them mock grades on their performance. Later, when his daughter was attending Radcliffe, I looked forward to the beginning and end of the year when he would drop in to my office to chat about geology, science, and people. These two men, one stormy, the other quiet, remade the field by their writings and through their students.

Twenty years before plate tectonics muscled aside all previous concepts about the origin of mountain belts, Krynine, Pettijohn, and their followers had set in place the general relationship between the large-scale movements of the Earth's crust and the kinds of sediment that these movements must produce. Knowing that genesis of sediment, we could draw maps of ancient times showing the long-vanished mountain ranges, the piles of sediment accumulated below the mountains, the paths of rivers that brought the debris of the mountains to deltas, and the shorelines of the oceans. We saw these paleogeographic maps in the framework of the times—a somewhat mysterious conglomeration of ideas on the dynamics of the Earth—and thought we knew with confidence what had happened and where, but could not, for all that, be sure how the whole thing was driven. We muttered and argued about the relationship of volcanoes to the evolution of mountain belts. We asked, for example, why so many—but not all—mountain belts lay along continental margins and whether continents had drifted. In hindsight, I imagine we were honing our tools, getting ready for the big event.

Geologic features and activities associated with plate collisions and subduction: ocean trenches, accretionary prism deposits, magmatic belts, volcanism, earthquakes (dots). The drawing is not to scale; the thickness of lithosphere is about 70 km, depth of the ocean trench 10 km, and the distance from trench to arc is 300 to 400 km.

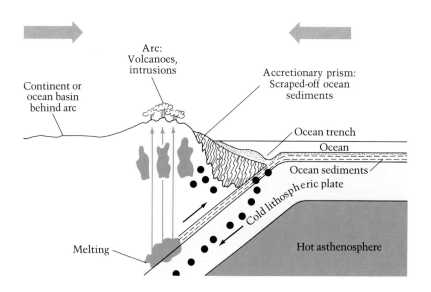

TECTONICS BECOMES
PLATE TECTONICS

When the theory of plate tectonics took center stage in the mid-1960s, it was quickly embraced by sedimentologists. It not only fit in perfectly with the general relationship between tectonics and sedimentation but illuminated that relationship, so that we could explore its corners. Now we could begin to think coherently about the distribution and composition of sands rich in volcanic rock fragments. Why were there so many fragments of volcanic explosions and lava flows deposited by some turbidity currents? What was the relation between other turbidity-current deposits and nonvolcanic compositions that seemed typical of detritus derived from the stable interiors of the continents? New research began to flood the journals, and provocative ideas were hurled about by young people relatively new to the field. Within a few years William Dickinson, a young sedimentologist from Stanford, proposed the relations that were to set the pattern for the next decade. Dickinson showed how we could link plate-tectonic settings—subduction zones, rift valleys, active continental margins, large transform faults, such as the San Andreas in California, and passive continental margins, such as the Atlantic Coast—to the mineral and rock-fragment ratios first explored by

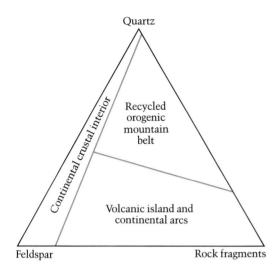

Quartz

Continental crustal interior

Recycled orogenic mountain belt

Volcanic island and continental arcs

Feldspar   Rock fragments

A quartz–feldspar–rock fragment composition diagram. Each point of the triangle represents 100 percent of that component; thus, a point at the center of the triangle would represent a composition of one-third of each component. Continental crustal rocks are rich in quartz and feldspar but poor in rock fragments; volcanic arcs are poor in quartz; recycled orogenic belts are rich in quartz.

an earlier generation (see the table of plate-tectonic settings on page 134).

Dickinson had done extensive research on the sandstones in California and Oregon and was familiar with sandstones rich in volcanic rock fragments found in tectonically deformed belts. Volcanic sandstones were a type that had been bypassed by many earlier sedimentologists, who had concentrated on the thin accumulations of stable sands of the midcontinent lowlands and the relatively gently folded sandstones that were remnants of older mountains such as the Appalachians. On the same triangular graphs that Krynine and Pettijohn had used to assess tectonism in general and to classify sandstones by type, Dickinson outlined broad regions of quartz, feldspar, and rock-fragment proportions that were associated with different plate-tectonic settings. For example, most volcanic-island arc detritus fell in a field distinct from the detritus of continental masses. Continental orogenic-belt materials were in yet another region of the diagram. Although these fields overlapped, it was remarkable to see how sands of the different tectonic environments clustered. Here was a tool for reconstructing a new kind of paleogeography, to fit the new global dynamic of plate motions.

The plate-tectonic interpretation of sedimentary rocks quickly became a three-ring circus. In one ring were the explorers of modern plate boundaries and their sedimentary environments. Jaunts over the subduction zones in the South Pacific became mandatory if we were to understand the dynamics of sediment transport on volcanic island arcs. The islands of an arc, the forearc region, and the deep-sea trench were surface expressions of the convergence of two oceanic plates (see the illustration on page 135). Sand deposition in modern rift valleys such as the East African rifts would be the guide to ancient valleys created by the tearing apart of a continent. In these and other settings sedimentologists sorted out the various sedimentary environments, from alluvial fans and plains on land to deep trenches beneath the ocean, and how they related to the motions of the 100-km-thick lithospheric slabs that collided, slipped past each other, and pulled apart.

In a second ring of the same circus were marine geologists, paleontologists, and geophysicists frenetically reinterpreting the geological history of the oceans in terms of plate tectonics. The raw materials for a new kind of geologic history were the mag-

Plate tectonic settings

Zone	Activity	Forms	Example
CONVERGENCE	Regions along the boundries of plates move against one another.		
Subduction	One converging plate, the oceanic subducting plate, descends into the mantle along a slope beneath the other.		
Intraoceanic island arc	Both plates are oceanic.	Trench, forearc, arc islands, marginal seas	Marianas Islands (Pacific)
Oceanic–continental (active continental margin)	The oceanic plate subducts beneath the continental plate.	Trench, forearc (land or sea), continental volcanic mountains, high plateaus, intermontane basins	Andes
Continent–continent collision	Both converging plates are continental; one may descend for some distance beneath the other, thickening the crust, or may move laterally along the plate boundry.	High mountains and plateaus (generally nonvolcanic)	Himalayas
DIVERGENCE	Plates move away from each other as new crust is created.		
Midocean ridge	The ocean floor separates as basalt wells up along central rift valleys.	Linear mountain ranges on sea floor	Mid-Atlantic Ridge
Continental rift valley	A plate boundry is created along faults as continental crust ruptures and two plates diverge. Narrow ocean gulfs may form along widening valleys.	Fault-bounded valleys on continents or narrow seas between continents	Red Sea and East African rift valleys
Passive continental margin	Continents drift apart as former rift valleys and gulfs widen to oceans; continental margins are no longer at plate boundries.	Broad continental shelves, slopes, rises; abyssal plains	Atlantic Ocean margins
TRANSFORM	Plates slide laterally past each other along major faults.	Vertical faults with horizontal motion	San Andreas Fault (California)

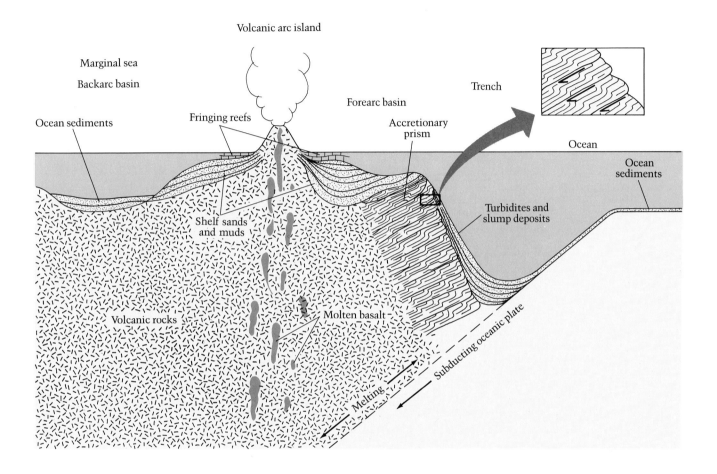

Volcanic arc island

Marginal sea

Backarc basin

Ocean sediments

Fringing reefs

Forearc basin

Trench

Accretionary prism

Ocean

Ocean sediments

Shelf sands and muds

Turbidites and slump deposits

Volcanic rocks

Molten basalt

Melting

Subducting oceanic plate

Sedimentary environments associated with intraoceanic subduction zones.

netic striping of the sea floor, parallel bands of similarly magnetized basalt; the fracture zones, transform faults originating at midocean ridges and extending hundreds of kilometers from the ridges; and the volcanic and sedimentary rocks of the sea floor. The evolution of the sea floor was to be read as a succession of changing plate configurations and relative motions. Plates were consumed and plates created, midocean ridges were swallowed by subduction zones, and the continents riding the plates were shifted around on the globe as the giant jigsaw puzzle constantly changed. The geologic timekeeper was the magnetic reversal timetable of the Earth, which flipped the north and south magnetic poles every few tens or hundreds of thousands of years and

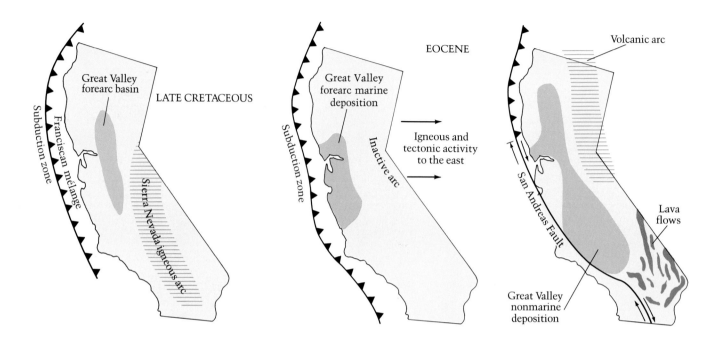

LATE CRETACEOUS

Subduction zone

Franciscan mélange

Great Valley forearc basin

Sierra Nevada igneous arc

EOCENE

Subduction zone

Great Valley forearc marine deposition

Inactive arc

Igneous and tectonic activity to the east

Volcanic arc

San Andreas Fault

Great Valley nonmarine deposition

Lava flows

The plate tectonic evolution of California since the Cretaceous period. An oceanic plate was being subducted along what is now the coast, producing the volcanic belt along the present Sierra Nevada Mountains, and a forearc basin where the Great Valley is now. *Left:* Late Cretaceous. *Center:* Eocene. *Right:* Pliocene, during which the subduction zone was partly converted to a transform fault, the San Andreas fault.

left a record of those reversals in the magnetic polarity of the basalt volcanics of the magnetic stripes on the sea floor.

The rocks of the sea floor were poked and prodded by the ocean-crossing drilling rig, the *Glomar Challenger.* The Deep Sea Drilling Program, started by a consortium of oceanographic institutions before plate tectonics had been proposed, was quickly pressed into service to confirm and then extend the new theory to the history of the oceans. As the core barrels of the *Challenger* brought up sample after sample of sea-floor turbidite, deep-sea limestone, volcanic ash, and all the rest, the scientific crew would plug them into a rapidly growing historic framework of world plate movements.

On land, in the third circus ring, stratigraphers, structural geologists, and sedimentologists had a harder job and a different set of forms, structures, and rocks to reinterpret. The magnetic striping produced by reversals of the magnetic field and recorded by basalt layers extruded at midocean ridges were absent. Instead of midocean ridges, geologists had mountain ranges; instead of fracture zones, extensive fault systems. Whereas the first blush of plate-tectonic theory seemed to work so well for the oceans, it was

The newest deep-sea drilling ship succeeded a similar forerunner, the *Glomar Challenger.* The new ship has a computer-driven positioning system that keeps it stationary while drilling from the central derrick.

more problematic on the continents. For the later geologic eras, the last 150 million years, the history of the oceans was a help. From that history we could get plate motions and then extrapolate what was happening on the sea floor to the adjacent parts of the continents. And surely, it seemed, the edges of the continents was where most of the action occured.

Subduction zones at continental margins led to distinctive kinds of mountain building. These volcanic arcs were not islands but high mountains bordering the continent. The Andes became a model for the complex of processes that resulted from the plunging of sea-floor lithosphere beneath a continental margin. As the surface of the descending slab of the eastern Pacific melted, it produced pods of molten rock that ascended to become the volcanoes and mountains of the western Andes, a volcanic arc on land.

The Earth's outer shell, the lithosphere, is divided into plates that move in response to convection currents in the mantle below. Their present-day relative motions are shown by arrows at plate boundaries.

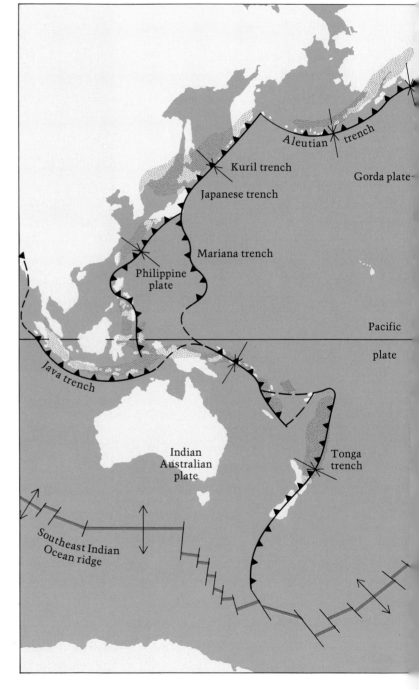

Divergent boundary

Convergent boundary

Uncertain plate boundary

Transform fault

Direction of plate motion   10 cm/year

Deep–focus earthquake belts

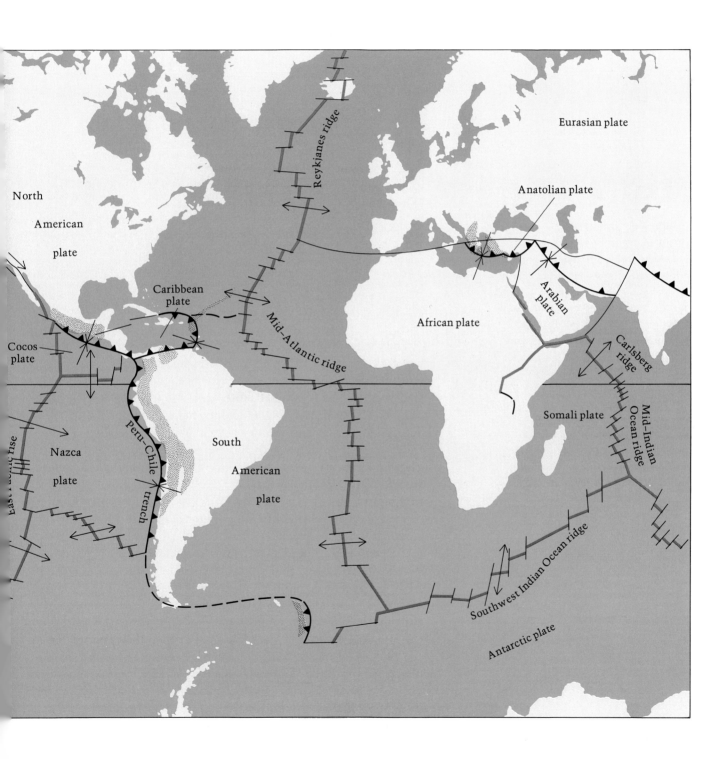

Right: The Peruvian Andes in a simplified cross section. Under the western Cordillera Blanca the crust has been thickened by intrusions of volcanic material rising above the Nazca plate as it plunges under South America. The convergence of the two plates also thickens the crust by pushing it together. Sedimentary environments range from marine shelf and trench over the forearc region to alluvial fans below the folds of the high mountains. *Opposite:* The formation of the Himalayas. Some 60 million years ago the oceanic lithosphere at the leading edge of the Indian plate was being subducted under southern Tibet *(A)*. Volcanic magma rising above the Indian plate formed granite intrusions. Sediments and oceanic crust scraped off the descending plate piled up in an accretionary wedge and forearc basin. Sometime between 55 and 40 million years ago the two landmasses collided *(B)*. A new fault, the main central thrust, broke through the Indian crust *(C)*. A slice of Indian crust, topped by Paleozoic and Mesozoic sediments, was thrust up onto the oncoming subcontinent. The forearc sediments were thrust northward onto Tibet. About 20 to 10 million years ago the Main central thrust became inactive. Since then India has slid northward along the Main boundary fault *(D)*. A second slice of crust has been thrust up onto the subcontinent, lifting the first slice. The uplifted slices make up the bulk of the Himalayas.

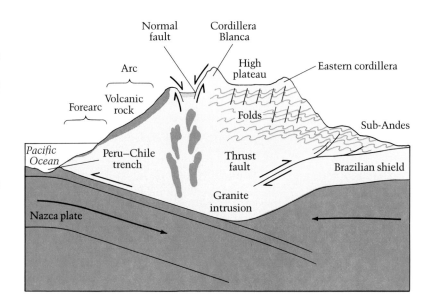

The highest mountains of the globe were found where two continents converged: the Himalayas became the prime example of a continental collision in progress after we realized that the Indian subcontinent had rammed into the Asian plate. Unlike the Andes, this mountain range had no volcanic arc but a thickened continental crust and faulted belts that raised the highest altitudes on Earth. Armed with these examples, geologists could revise the historical record of the other great mountain belts of the world. Naturally the origin of the Alps, the first of the great mountain chains to be explored in detail by generations of European geologists, became the first order of business. Here were the classical flysch and molasse, the great overthrust faults, and other features that we could now see as the hallmarks of collisional mountain building. The conceptual framework was there, all right, but the adjacent ocean that would give the clue to the plate history was missing. The Mediterranean was itself too complex and gave a record only of microplates—splinters and fractures of continental and oceanic crust far smaller than the major plates—and the more recent history, sedimentary and tectonic, of the region. So the modern alpine geologists were on their own, which put them in the same fix in which geologists studying much older mountains found themselves. These geologists had to deduce plate motions from rocks and structures alone.

A

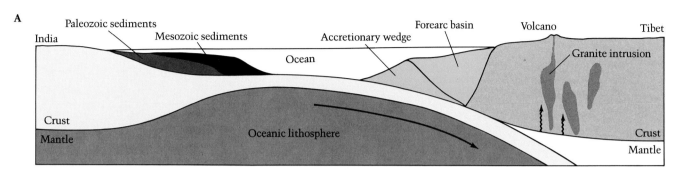

India · Paleozoic sediments · Mesozoic sediments · Ocean · Accretionary wedge · Forearc basin · Volcano · Tibet · Granite intrusion · Crust · Mantle · Oceanic lithosphere · Crust · Mantle

B

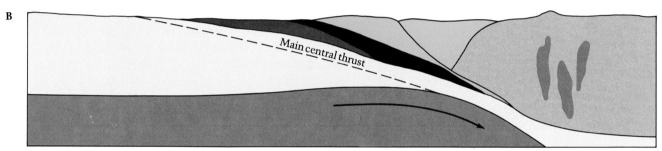

Main central thrust

C

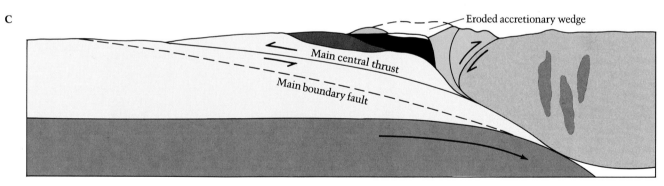

Eroded accretionary wedge · Main central thrust · Main boundary fault

D

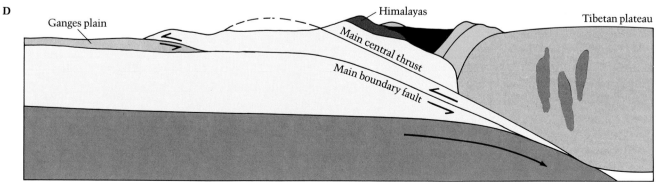

Ganges plain · Himalayas · Tibetan plateau · Main central thrust · Main boundary fault

For example, knowing that the oldest sea floor now preserved intact is Jurassic in age (a little less than 200 million years old) and found only at the edges of the major oceanic plates was of no help either to alpine geologists or to geologists in the eastern parts of North America who wanted to know the history of the Appalachian Mountain chain. Extending in a great looping path from the Marathon region of west Texas to the Maritime Provinces of Canada, the Appalachians always posed a challenge to those who would explain mountains. Geologists incorporating continental drift could now pick it up again in East Greenland, in the Scottish Highlands, and as far as the mountains of Norway and Sweden.

Geologists would do historical geophysics through historical geology. And of all the rocks—of the Alps, the Appalachians, and other mountain belts—they would study the sediments, and of these, the sandstones would be central to understanding the sequence of orogenic events. The sandstones could reveal the source areas, paleocurrents, and sedimentary environments in which changing configurations of plates and plate boundaries interacted. The geologists were able to link plate-tectonic settings of the present and recent past with the rocks and structures of more-ancient mountain belts and undeformed continental interiors.

PLATE-TECTONIC SETTINGS

Geophysics provides the dynamic backdrop that explains how the geology of a region relates to its tectonic origins. Lithospheric plates are created at midocean ridges, move about over the surface of the Earth, and dive into the interior at subduction zones, all in response to large-scale convection currents in the mantle of the Earth. Mid-ocean ridges form where rising mantle convection currents cause the upwelling of molten basalt. The melted basalt congeals after it works its way up volcanic vents and fissures to the surface, where it quickly cools. The injection of masses of basalt is a central feature of sea-floor spreading. This dynamic of ocean-floor rifting and emplacement of volcanic rock determines the sea-floor environment of the ridge. A central rift valley dotted with steaming vents that spew water and gas at temperatures up

The Appalachian Mountains extend from central Alabama in the southern United States to Newfoundland in southern Canada.

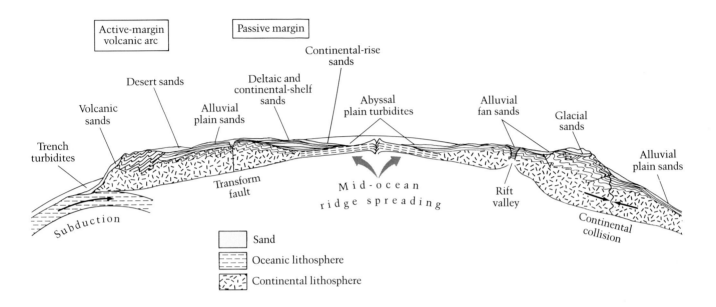

Active-margin volcanic arc

Passive margin

Continental-rise sands

Desert sands

Deltaic and continental-shelf sands

Volcanic sands

Alluvial plain sands

Abyssal plain turbidites

Alluvial fan sands

Glacial sands

Trench turbidites

Alluvial plain sands

Transform fault

Mid-ocean ridge spreading

Rift valley

Subduction

Continental collision

Sand

Oceanic lithosphere

Continental lithosphere

Plate-tectonic settings of sedimentary environments in which sand is deposited.

to 350°C bisects a chain of high submarine basalt mountains.

Sinking motions of the mantle drive the subduction of oceanic lithospheric slabs. As one slab slides beneath the other, the descending slab melts, provoking volcanic activity and creating a volcanic arc. The geological environments pass from the deep-sea trench, formed where the slab starts to descend, to the arc of volcanoes. Between is the forearc region where relatively shallow aprons of sediment accumulate. Where plates slip past each other horizontally, great fault systems such as the San Andreas of California tear apart the continent and displace mountains and valleys. Slippage along curves produces pull-apart basins. Rifting of continents is the response to tension along zones where the mantle may be upwelling and diverging. The rift valleys are distinguished by their broad, low valley floors, steep walls, and volcanoes or fissures emitting basaltic or other lavas. Ultimately, these components of the global pattern of plate and mantle convergences and divergences are responsible for the properties of the sands formed in the sedimentary environments of the Earth. Far removed from, but still connected to, those ultimate causes, the sandstones that we read in the field and laboratory tell us the bounds of the natural world, tell us which events actually happened in its history.

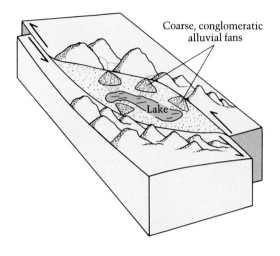

Coarse, conglomeratic
alluvial fans

Lake

The formation of a sedimentary basin along a transform fault. A curve in the fault may pull apart adjacent sections as the fault moves. These pull-apart basins are filled with sand, gravel, and mud weathered from the adjacent highlands.

SUBDUCTION ZONES: INTRAOCEANIC

Deep-ocean trenches in the southwest Pacific Ocean have piqued our curiosity ever since they were discovered by H.M.S. *Challenger*, the pioneer oceanographic ship, in 1872. The deepest of them, Tonga–Kermadec lies over 35,000 feet below sea level, over a mile deeper than Mount Everest is high. Formerly a durable mystery, these trenches seemed to oceanographers to be possible forerunners of geosynclines, linear belts of great thicknesses of sedimentary rocks that were ultimately deformed into mountain chains. Yet the trenches seemed too devoid of sediment for that to be true: they pointed instead to some deep-seated forces that pulled them down. To use a favorite word of the time, the normally bouyant oceanic crust here "foundered." But no one could specify the forces or spell out the dynamics until the advent of plate tectonics. The new model explained the trenches: they were the surface manifestation of the descent of one oceanic lithospheric plate below another.

The Pacific volcanic island arcs are the reverse image of the trenches: in acting out the subduction dynamic, "What goes down must come up" is the apt statement; the material at the surface of the descending slab melts and works its way to the surface because it is less dense. Between the arc and the trench is the forearc region. At the trench side of the forearc, some of the thin sediment of the sea floor is scraped off and stacked in a series of faults, plastering it onto the forearc in a prism of deformed sedimentary rocks. So it turns out that sediment does accumulate, but mainly at the side of the trench, not at the bottom. Between the deformed, faulted mélange, as it is called, of oceanic sea-floor scrapings and the arc itself is a shallow forearc basin that aprons the arc. In the many island arcs that lie in the tropics this basin is the site of coral reefs and shallow platforms that accumulate the calcium-carbonate debris of thousands of shelled organisms.

The islands of the arcs of Tonga–Kermadec are small, and erosion of these heavily vegetated, coral-fringed, volcanic islands produces mainly clays. The only source of sand-grain-sized fragments is the volcanic material that spews into the air or cascades

The ocean floor in the southwestern Pacific Ocean, showing the many trenches and other forms of plate tectonics.

down the volcanoes' slopes, sometimes under the sea. Outside the tropics, volcanic islands of an arc such as the Aleutians, have ocean water too cold to support fringing coral reefs and the great hunters of shelled organisms living on the reefs, and carbonate sediment is absent. At the same time, mechanical weathering of the bare-rocked volcanic islands is intense in the frigid, stormy climate of the Bering Sea. Volcanic rock fragments eroded from older island volcanics mix in abundance with fresh contributions from active volcanoes. In exploring these contrasts in weathering, we have uncovered a climate indicator in the sediments of island arc regions, linked on the one hand to carbonates and on the other to coarse weathering products. Such climatic differences can be read in the sediments of ancient as well as modern island arcs, and we can check the latitudinal differences implied by paleomagnetic measurements that give us the continental-drift paths of the islands.

Cold and icy or warm and humid, the oceanic island arcs are not prolific producers of sediment, at least not compared with a continent's flow of detritus. Because much of the volcanic component is ejected as fine-grained ash, much of the sediment is muddy rather than sandy. The sediments from the descending slab are typically open-ocean, or pelagic, sediments made up of fine clays and, where surface waters are favorable for growth of pelagic organisms, chemical components such as calcium carbonate or silica. In the long journey that the older parts of the Pacific plate will have made before they disappear into a subduction zone, they accumulate these sediments slowly. At rates of sedimentation as low as a few millimeters in 1000 years, the sedimentary layers of the open ocean are only a thin, creamy frosting on the basaltic lithosphere. Pity the poor sandstone sedimentologists—they have no reason to spend arduous months of field work on the palm-covered, coral-beached oceanic islands of Tonga or the Marianas. Only the largest arc islands, those of the Philippines or Japan, produce quantities of sand.

The size of a volcanic island arc relates to its long, varied history. As oceanic arcs evolve through millions of years of continued subduction, the volcanic cones that build up on the surface are matched by an "underplating" of igneous rocks that congeal at depth—growing masses of coarsely crystalline rocks from magmas that do not make it to the surface. Sediments accumulate from the weathering of the igneous rocks and are buried. New volcanic cones break out at random and disturb older masses. Buried sediments and igneous rocks are baked and pressurized, creating suites of metamorphic rocks whose compositions and textures obscure their origins. Imperceptibly, the island arc grows to a small massif that begins to look a little bit like a continent. The islands of Japan have had this kind of history over hundreds of millions of years, one volcanic-dominated terrane giving way to the next, younger one. The very busy geologic map of Japan reflects the details of a history of changing plate motions, the subduction of younger and older ocean floors, oceanic plateaus and ridges, and to make the mix a little more exotic, some continental slivers from the Chinese mainland.

The sandstones of Japan provide an important clue to these events. Ultimately they all are volcanically derived, their compositions a guide to the source areas that were being eroded at the time. Grains of older volcanic rocks lie cheek-by-jowl with shards of fresh volcanic glass. Grains whose volcanic origin is still recognizable are now metamorphic rock fragments. Igneous rocks of differing compositions abound. Pure quartz sandstone, however, could not be found in any of the Japanese samples.

Not surprisingly then, when I was lecturing at the University of Tokyo, some years ago, there was excitement among my Japanese hosts as they introduced me to the young geologist who had found the first quartz sandstone in Japan—a Precambrian one at that. The excitement was a bit controlled and I soon discovered why. This alert young sedimentologist had noticed not a whole new formation, except by distant implication, but a single pebble in a conglomerate. The discovery, however tiny, did imply that somewhere at that time, this little piece of the Japanese islands had received some debris from a continental source, perhaps when this sliver of land was part of a mainland in the cloudy past. Found in the midst of a volcanic arc massif, a single pebble of that composition could only have come from a place where quartz-rich rocks weathered so deeply that all minerals other than quartz had disappeared in their trial by weathering, erosion, and transportation. Characteristic of oceanic volcanics, the Japanese rocks contain little quartz: intense weathering could not have left enough to form an exclusively quartz sandy formation.

If Japanese sandstones are largely volcanic, they are nevertheless heterogeneous in their sedimentary structures, their layered sequences, and other properties we use to infer sedimentary environments. Many are turbidites, signaling their deposition in deeper-water troughs bordering the islands. Some carry the marks of ocean trenches: packets of graded sandstones and shales disrupted by blocks of slumped material that slid down the trench wall. Others show evidence of deposition on shallow marine shelves bordering the islands, much like the shallow waters between the islands or the narrow shelves bordering the Sea of Japan today. We find river-deposited sands interfingered with volcanic-ash falls and flows. If we integrate the paleogeographic mapping of Japanese sedimentologists with the history of the sea floor of the northwest Pacific, which shows shifting scenes of subduction, we can get a good picture of the evolution of this complex island arc for the last 200 million years. We can rely only on rock mapping for the terranes older than that but still can do remarkably well for the period between 300 and 200 million years ago. Before that, the history is shrouded, a little like the story of the earliest inhabitants of the Japanese islands 3000 years ago, the Haniwa people. We have relics and fragments but no narrative.

SUBDUCTION ZONES: OCEANIC–CONTINENTAL

The trenches of the western Pacific mark the descent of oceanic lithospheric slabs beneath volcanic arcs on other oceanic slabs or, as in the case of Japan or the Philippines, beneath oceanic slabs with arc massifs that begin to look something like tiny continents. Off the western coasts of the Americas the story is different. The Peru–Chile trench marks the subduction of the Nazca plate under the South American continent; the middle Americas trench starts the Cocos plate beneath Central America; and to the north, minus a deep trench—more on that soon—the Juan de Fuca plate slips below Oregon, Washington, and British Columbia.

   Two consequences of subduction beneath continents make the geology of continent–ocean convergences strikingly different from intraoceanic convergence zones. The first creates some of the major topographic features of continental margins bordering subduction zones. When the melting at the surface of the subducting slab takes place under a continent, the result is a chain of volcanoes making high mountains. The Andes, the central American coast ranges, and the Cascades of the Pacific Northwest are all arcs on land. The arc may be some distance from the trench because the angle of subduction is less under continental than under oceanic crusts—a fact that geophysicists argue about but cannot yet explain. The second consequence of continent–ocean subduction is an abundant sediment supply to lowlands below the rapidly eroding arc mountains. Gravels, sands, and muds pour onto the coastal forearc region and may be transported over the marine shelf to the trench. Billions of tons of sediment may be

The Columbia and Snake rivers carry a mixture of detritus from the Rocky Mountains, the basalt plateaus to the west, and the Cascade range.

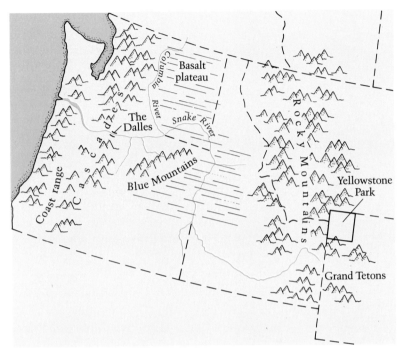

dumped at the continental side of the arc mountains and build up a complex sequence of sediments on the foothills and plains below the mountains.

Compared with some others, the Juan de Fuca trench seems insignificantly deep because the Columbia River has filled it with so much sediment from the continent. The Columbia carries a huge sediment load derived from different terranes. In its headwaters in the Canadian Rockies and in the headwaters of its major tributary, the Snake River in the U. S. Rockies, mineral and rock fragments of a major continental mountain chain are the main sediment load. Here are quartz and feldspar, metamorphic rock fragments, and all the other components of the granites, sedimentary rocks, and metamorphic belts that are welded into the North American Cordillera. Descending to a lower plateau in eastern Oregon and Washington, the Snake and the Columbia enter a great basalt province. The rivers cut through what seems to be a sea of layered lava flows, with only a few peaks of older mountains—the Blue Mountains of eastern Oregon—sticking their fingers up through the volcanics. These "flood" basalts welled up through great fissures in the Earth's crust from a rising plume of hot mantle hundreds of kilometers below the surface.

River sands from the Rockies are nearly disguised by the new load of sand weathered from the plateau basalts, but patient work at the microscope still reveals the grains' multiple parentage. The Columbia then enters The Dalles, the beginning of a narrow gorge that cuts through the highest parts of the Cascade Mountains. The gorge was cut by the river as the Cascades were thrown up by incessant volcanism—from the subduction of the Juan de Fuca plate. Mount Saint Helens is only the most recent of a series of eruptions that built all the famous peaks, from Mount Hood to Mount Rainier and Mount Baker. As the volcanic chain grew, the river cut through it, adding a new kind of volcanic rock fragment to the river bed: andesitic rocks—the name bespeaks precisely similar rocks of the Andes. Andesitic rocks of the Cascades are distinguishable from the plateau basalts because of mineralogical and chemical differences in the water and other components of sediments that are added to the molten mantle as the subducting slab carries with it some of its sea-floor veneer. Our inventory is getting impressive: the river now carries continental-mountain-chain sand, flood-basalt fragments, and andesite-arc grains.

Dropping down from the Cascades, the Columbia crosses the fertile Willamette valley and gets a shot of clayey Cascade detritus altered by the wet temperate climate of the lowlands. Not through yet and now a great broad tidal estuary, the river takes a final lap through the Coast Range, a modest line of turbidite sandstone mountains. We see cannibalism here, for these turbidites were formed at an earlier stage of Juan de Fuca subduction. Now uplifted, these former submarine-canyon and continental-rise sediments are made up of many of the same components as the modern sands of the Columbia River. At the end of its course, carrying its full complement of sand and mud, the Columbia reaches the sea and starts to drop its load. Strong longshore and tidal currents spread the far-traveled detritus over a narrow continental shelf and then into deeper waters, where its constantly fills what would otherwise be the deep depression of a trench.

If we knew nothing about the modern paths of the Columbia and the Snake, the current geology of the Pacific Northwest, and the geophysics of the sea floor and the Juan de Fuca plate, could we reconstruct them? A double-blind test has not yet been planned, maybe partly because any geologist versed in the required methods would probably be at least vaguely familiar with the lineaments of North America. But we can be confident at least that we could decipher the kinds of source terranes if not their precise positions. Correlating the sedimentological information with the geological history of the region, already known from mapping of the earlier continental rocks and structures, would then provide the general lay of the land.

That geological history might be as different from the history of another continental arc region as one Western European country's history is from another's. The Andes are a double chain of mountains, and only the Western Cordillera is active volcanically. The Altiplano, one of the highest plateaus of the world at 12,000 ft, is a backarc basin receiving millions of tons of sand, gravel, and mud from the peaks of the Eastern and Western Cordillera. The volcanic arc of Central America is yet another story, complicated by the Caribbean plate on the Atlantic side. As different as they are, all oceanic–continental subductions share the dynamic of the subduction that connects andesitic volcanism, mountain building, erosion, and sedimentation on the continent and offshore. A different dynamic takes over when subduction changes to continental collision.

CONTINENTS IN COLLISION

The Persian Gulf holds obvious interest for geologists as well as for oil economists and the war departments of the Middle East and the major powers. A subduction zone under the gulf is converging the Arabian plate with the Anatolian plate, which runs through Turkey and Iran. For most of this plate boundary, the subduction is of the Gulf's oceanic crust dipping below the coast of Iran. Very soon, however, continued subduction will close the gulf and smash the Arabian landmass against Iran. When that happens, the present Zagros Mountains of western Iran, now a mere 2.5 miles high, will grow even higher, in a pattern that repeats itself throughout the history of mountain belts.

The major high mountains of the world are the result of the collision of two continental masses. The most recent are the Himalayas, thrown up, as we have seen, as the Indian subcontinent moved north against the Asian plate. Only a little less recent are the Alps, the product of Europe's convergence with a southern plate of the time. Going back in time we next meet the Laramide orogeny of the North American Cordillera, formed when North America collided with former continental plates of the eastern Pacific. Before that came the collision between North America, Europe, Africa, and South America that created the Appalachian chain as part of the giant belts of mountains of the supercontinent Pangaea.

Mountains that we now recognize as products of continental collisions were centerpieces of geological theories in the nineteenth century, classics to be studied and examples that illustrated tectonic control of sedimentation. Now they have become the neoclassics, challenging new generations of plate tectonicists and sedimentologists to model their histories in terms of changing plate motions and geological consequences. Although each mountain chain has a unique set of successive events, a general pattern emerges: all show evidence of early volcanism and rocks and structures that spell subduction, either intraoceanic, oceanic–continental, or the first followed by the second. After a period typically lasting tens of millions of years, a continent—or the islands of a grown arc massif—that is riding along on the subducting plate meets a continent on the overriding plate. Continued convergence of two bouyant continental crusted plates thickens the crust at the convergence, produces complex, thrust-faulted mountain belts, and forces lateral squeezing of continental and subcontinental masses. When subduction turns into collisional orogeny, the volcanic arc stage is over, and volcanics cease to be the identifying process.

In all these orogenies, the erosion of the newly formed mountains sheds a great load of sandy debris on both sides of the chain. The sedimentological microscopist can see in these sand grains the minerals that made up the source rocks of the mountains: they are continental rocks rich in quartz and the feldspars, which are rich in sodium and potassium. Volcanic rock fragments are few and likely to have been metamorphosed to schists and slates. Sedimentary rock fragments may dominate as erosion strips the upper parts of the crust. With continuing uplift and erosion, deeper parts of the crust are exposed, and abundant sedimentary rock fragments give way to metamorphic rock fragments mixed

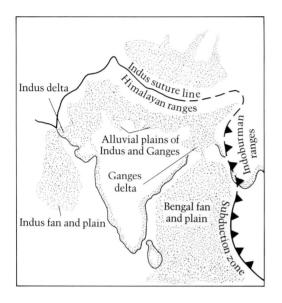

The huge load of sediment eroded from the Himalayas is carried by the Ganges and Indus rivers to their deltas, and further transported under the sea to submarine fans and plains.

with the mineral grains that make up granitic rocks. Just as complex as the Columbia River mixture, the mixture of grains that travels from the Himalayas along the Ganges River is also diagnostic of its origins. We can read the continental collision of the Himalayas from Ganges sediment, just as we could the ocean—continent subduction from the Columbia. The types of sedimentary environments associated with collisions, however, may be less distinctive than the source areas. The sedimentary environments of the Himalayas and the Andes are similar: rivers, intermontane basins, coastal plains, deltas, and offshore shelves and slopes.

The distribution of sediment eroded from the Himalayas shows how varied are the environments of deposition. The sands of the Himalayas accumulate first at the base of the mountains, in alluvial fans and plains of the Indus, the Brahmaputra, and the Ganges. Sluicing down toward the sea, the rivers leave a wide swath of valley fill many kilometers thick. At the entrance to the sea, at the head of the Bay of Bengal, the Ganges breaks up into dozens of distributaries and forms a great delta. Beyond the delta, currents carry sands and silts out to the edge of the continental shelf and then to the depths of the Bay of Bengal, where they accumulate as great fans of turbidites. Here is a paradox: tectonically controlled sedimentation in which flysch and molasse are contemporary, and formed at different stages of mountain building. Only if we look at the mineral composition can we differentiate between volcanic-rich flysch and continental block flysch. Geologists differentiate between flysch and molasse in the Alps and elsewhere by the bedding, the grain size, the sedimentary structures and the bedding sequences and those are impressed by the environment, not the geologic stage of mountain building.

The scale of the Himalayas and its sediment distribution pattern led to a provocative comparison. William Dickinson and two of his students, Stephen Graham and Raymond Ingersoll, superimposed the map in India and the Bay of Bengal on a map of eastern and central North America showing the distribution of the major Paleozoic mountain belts and their sedimentary facies. They suggested that the plate convergence that formed the main part of the Appalachians in later Paleozoic time shed sediment in somewhat the same kind of pattern, with the land-deposited sands lying along the present areas of the Appalachians and the

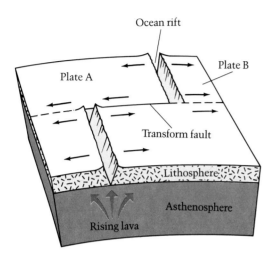

Ocean rift

Plate B

Plate A

Transform fault

Lithosphere

Asthenosphere

Rising lava

Transform faults along midocean ridges are fracture zones along which plates slip horizontally between offset spreading segments of the ridge.

turbidite marine equivalents along the present Ouachita Mountains of Oklahoma. In this business, we can stay tethered no longer to a large region than to a small field area: we have to look at a large part of an entire continent to see the gross pattern.

TRANSFORM FAULTS

Plate convergence and divergence, particularly as these affected the oceanic lithosphere, were the first plate motions to be harnessed to the general theory of plate tectonics. Hot on their heels came recognition of the lateral slippage of one plate along its boundary with another: transform faults, long lines of faults whose dominant motion is horizontal, are the surface expression of such plate boundaries. These faults were first mapped as offsets of midocean ridges, some of them fracture zones hundreds of kilometers long. Transform faults on land make up some of the major fault systems of the Earth's continents. The San Andreas, the best-known fault in the world, is legendary for having caused the San Francisco earthquake of 1906 and is the fault that worries city councils, homeowners, and real estate dealers as well as seismologists trying to predict its movements. As the Pacific plate creaks to the northwest past the North American plate, the San Andreas tears now along one segment, now along another.

Where the fault takes a turn—faults rarely run in perfectly straight lines—it may pull apart a chasm in the crust. Walled with steep cliffs, the newly created hole becomes a sediment basin for the detritus eroded from its borders. John Crowell of the University of California at Santa Barbara has mapped the Ridge basin, one of these pull-aparts along the San Andreas fault system that developed in the Pliocene epoch, about 5 millions years ago. Earlier, working with Kuenen, Crowell was one of the first to recognize the alpine flysch as a series of turbidites. Combining his talents as a structural geologist with his interest in sediments led him to map the composition and distribution of sediment types in relation to the evolution of the basin, which stretched open as the faulting progressed. Conglomerates and sandstones many kilometers thick—some of them alluvial fans, others turbidites deposited in an arm of the sea—filled the basin at the faulted margins. With distance from the fault, the sediment quickly became finer-grained and merged with much thinner lake

The San Andreas Fault, a very long transform fault along which the Pacific plate slips north and west past the North American plate.

deposits. The rock and mineral fragments of these sediments clearly betray their origins from the rocks of the walls around them. They are thoroughly continental, and their large size and lack of sorting are evidence of brief transport. Such basins have been recognized in many other areas; one of the spectacular sights of the world, the Dead Sea, occupies a depression produced by the northern movement of the Arabian block on the east against the Palestinian block to the west.

PLATE DIVERGENCE: MIDOCEAN RIDGES AND CONTINENTAL RIFTS

If plate convergence creates mountain chains, what does divergence do? A symmetrical process would create an inverse mountain chain—an elongate depression in the crust—and that is more or less what happens. Along the central axis of a midocean ridge we find a deep, narrow valley, recognized in the 1950s, the decade before plate tectonics, as a rift produced by the separation of crust under tension. Later seen as a natural consequence of sea-floor spreading, these valleys are the site of extensive faulting produced by the tearing apart of the crust as basalts from the mantle well up from a rising convection current. Oceanic rift valleys are hardly places of extensive sedimentation, certainly not of such coarse material as the sand from eroded mountains. Yet we find

Cross section of the central rift valley of the Mid-Atlantic ridge in the FAMOUS (French–American Mid–Ocean Undersea Study) area southwest of the Azores. The downfaulted deep valley is the site of most basalt lava flows.

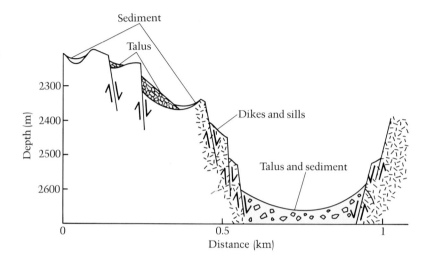

these depressions high on submarine mountains rising to great heights above the low abyssal plains of the deep-ocean basins. Height matters little in the minimal erosion that occurs under seawater. In contrast to the rain, wind, ice, and beating sun of the land, these rocks stay at the constant cold temperatures of the ocean floor, close to 0°C, where chemical weathering is extremely slow. Without the physical fragmentation caused by ice freezing in cracks, plant root growth, and weather weakening of cracks, the basalts of the ocean floor produce little detritus. The valleys are unfilled in their transitory life as the sea floor forms and then spreads to become part of the uplifted ridge flanks.

We do find sand of an unusual kind as sediment on midocean ridges. Composed entirely of calcium-carbonate shells of unicellular foraminiferans that live in surface waters, these calcareous "oozes" drape the crests of the ridge mountains as the shells settle to the sea floor after the organisms die. There they collect until one of the numerous earthquakes generated by tensional faulting triggers a turbidity flow down the slopes to the valleys and basins below. As they flow down, they may incorporate a few fragments of basalt from the bedrock of the slopes; in this they differ from the oozes that cover large areas of the sea floor where the topography is gentler. Only where midocean ridges come close to land, as does the Juan de Fuca in the Pacific Northwest, is

As continental crust is stretched, it thins and falls apart by faults, creating a rift valley that becomes filled with sediment.

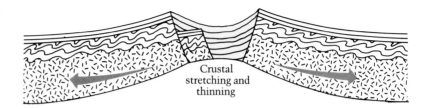

Crustal stretching and thinning

sediment derived from the land a significant part of the ridge province.

To sandstone sedimentologists, rift valleys on the continents are much more fascinating than midocean rifts. Despite the excitement of being one of the few to observe a midocean ridge from a submersible, there is stronger recompense for doing field work in a continental rift valley. The geologist there can see both the detail of small areas and the overall picture of the entire rift system. Sampling is easy and the variety of continental rocks great. In a currently active rift, the geologist can compare the rocks of the continental borders of the rift with the sediments of the valley derived from them. The sediments derived from the borderlands may be mixed with those from the erosion of volcanics extruded along the valley. Sands, gravels, and muds are interfingered with lake deposits common in the lower parts of the valley.

Rifts hold great fascination for yet another reason. The genus *Homo* picked the African rift for a place to evolve. Our progenitors lived along the rivers of its valleys, swam and fished in the lake waters, and were occasionally buried in the volcanics of the rift. Here, in the river and lake sediments of Olduvai Gorge, Koobi Fora, and the other now famous places of the East African rift, Louis Leakey, his wife, Mary, and his son Richard found the bones that mark the evolution from *Homo erectus* to modern *Homo sapiens.* Geologists, anthropologists, and paleontologists have scoured the region for more bones, worked out the geology of the rift to give the historical framework on which to hang hominid evolution, and analyzed the sedimentary environments in which our ancestors lived. Those habitats were influenced by the climate as well as the tectonics. Sometime around 5 to 7 million years ago, when the apes and the hominids began to diverge, the Earth's climate began to cool and the glacial age was upon us. The

Dinosaur tracks in sandstone.

woodland, plains, and grassland environments associated with the rift valleys were the product of climate interacting with tectonics in a way that was peculiarly favorable for this—to us very special—evolutionary path.

The sands of the African rift reflect this set of environments. The materials are recognizably derived in large part from continental crustal material from the borderlands, rich in quartz, feldspar, and sedimentary and other rock fragments. Mixed with this detritus are volcanic fragments ejected by Kilimanjaro and other volcanoes. Where and when climates were hot and arid, the chemical deposits of alkaline lakes—borates, silicates, and carbonates—were precipitated. At places and times when fresh waters dominated, shrinkage and expansion of the lakes in response to climate could be correlated with the interfingering of lake and river sediment. We do not yet know how the primate and hominid populations were affected by climatic changes superimposed on the unsteady faulting and volcanism of the rift. It is an open question whether these changes simply accompanied the evolution or might have forced it.

Sands like those of the East African rift are well known from other times and places. In the Triassic period, a little over 200 million years ago, a giant rift system fractured the supercontinent of Pangaea roughly along the lines where Pangaea had been assembled from the Paleozoic continents. At first, these fractures were continental rifts, much like those of East Africa. The sandstones of the Newark Series, as they are known along the eastern seaboard of North America, were deposited in the same general system of rifts as the famous New Red sandstones of Britain. Sandstone bearing the marks of origin as alluvial fans, alluvial plains, freshwater lakes, and piles of basaltic lava reveal the trace of the rift. These valley environments were the habitat of small dinosaurs. Along the Connecticut River, which now runs over the former rift, are outcrops of the alluvial sandstones that record that time. Students seeing these outcrops for the first time are often wary, but then, exhilarated with recognition, jump along the clear footprints of the small, running reptiles that lived on these ancient river shores. For small dinosaurs, too, a rift valley was a good place to live.

Some rifts recorded in sandstones, for example, those far to the east of the rifts of the Newark Series, buried deep beneath the

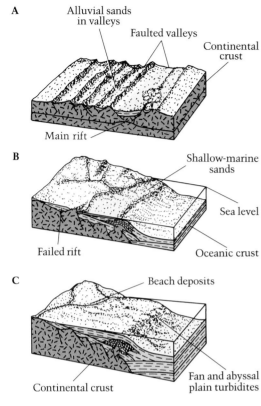

A — Alluvial sands in valleys — Faulted valleys — Continental crust — Main rift

B — Shallow-marine sands — Sea level — Failed rift — Oceanic crust

C — Beach deposits — Continental crust — Fan and abyssal plain turbidites

D — Coastal-plain sands — Continental shelf — Continental slope — Continental rise — 0 km — 7 km

The development of a passive continental margin along the rifted Atlantic coast of the United States. A rift zone develops as an ancient continent stretches and thins. Volcanics and Triassic nonmarine sands and clays are deposited in the faulted valleys *(A)*. Sea-floor spreading begins along the main rift; the lithosphere cools and contracts under the receding continental margins, which subside below sea level; subsidiary rifts of the zone become failed rifts and continue to receive sediment. Deltaic and other deposits *(B)* are deposited and then covered by Jurassic and Cretaceous sediments derived from continental erosion *(C and D)*. The Atlantic margins of Europe, Africa, and South America have similar histories.

sediments of the Atlantic continental shelf, are detectable only by geophysical methods. These rifts were at first intracontinental like their more western counterparts, but they soon developed into genuine plate divergences. As Pangaea split, narrow oceans invaded the rift valleys, and midocean ridges formed. As spreading continued, the early Atlantic Ocean formed, just as the Red Sea today is spreading and dividing Africa from the Arabian plate. The total rift zone was hundreds of kilometers wide, extending from failed rifts on the landward side—rifts that did not ever widen sufficiently to part the crust of the continent and allow basaltic oceanic crust to intervene—to the successful seaward rifts that did pull apart to form an ocean. The Newark Series failed rifts had an uneventful history after their geological moment of furious activity, but other failed rifts continued to play an important role in continental evolution.

The low, rolling topography of southern Illinois, broken only by an occasional ledge of sandstone and a higher hill or two, masks a sedimentary basin underlying much of southern Illinois and adjacent parts of western Kentucky, Missouri, and southwestern Indiana. That basin started as a failed rift. Beneath the surface is a spoon-shaped depression in the crust of the Earth that holds a series of late Precambrian and Paleozoic sedimentary rocks up to 12 km thick. Geophysical evidence, deep seismic sounding of the crust by artificial impulses of compressional waves, like some of those produced by earthquakes, now leads us to believe that the floor of the basin is an ancient failed rift. A few very deep drill holes show the presence of feldspar-rich sands that filled this basin early in its history. These rifts were active in the late Precambrian, a time when events took a turn similar to the much later breakup of Pangaea in the Triassic period. About 1 billion years ago a supercontinent rifted apart to form an ocean called by geologists Iapetus, in Greek mythology one of the Titans and the father of Atlas. Iapetus formed off the east coast of the continent but left behind a trail of failed rifts, some of them, like the Illinois basin, in the central part of the continent.

Confirmation of this rift and exploration of the early events that led to the Illinois basin may come from a new program of deep continental drilling sponsored by the National Science Foundation (NSF). The Illinois State Geological Survey and a consortium of geologists has proposed to the NSF that a superdeep

Map of failed rift valleys beneath the Paleozoic sedimentary rocks of Illinois and adjacent states. These late Precambrian rifts never spread to form a new ocean but did subside and initiate sedimentary basins.

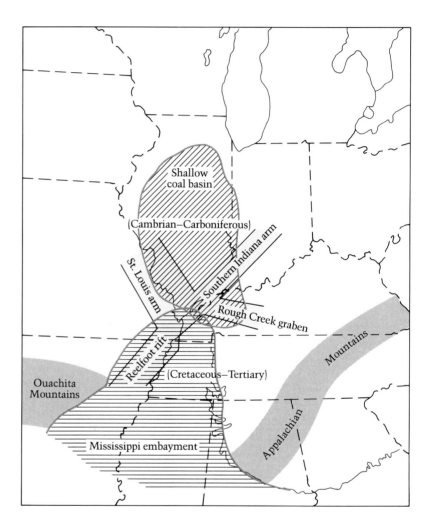

hole, 12 km down, be drilled into the rift floor of the basin. If and when the hole is drilled, geologists will be able to sample the sandstones that filled the rift early in its history, sands derived from a continental surface a billion years ago. Exploration of this stage of the basin's history is part of the search for a mechanism that will explain how the rift continued its intermittent activity throughout the Paleozoic Era, while deeply buried by later sediments.

The historical crustal lineaments of this rift zone are still active. A devastating North American earthquake struck the town of New Madrid, Missouri, in 1812, after movement occurred along a fault deeply buried in the rift system under the Illinois basin. A more recent quake in the Illinois basin in 1987 points up the continuing activity of the faults in the crust deep beneath the flat-lying sedimentary surface rocks. A story that started with the East African rift has led us to earthquakes and a billion-year-old continental separation. One of the rewards of sandstone geology is the vast canvas of geological processes and histories that opens up to us when we trace out a single line of inquiry on a set of sand grains.

PASSIVE CONTINENTAL MARGINS

Some of the world's most prolific oil fields are on the Gulf of Mexico coast in Louisiana and Texas. The Tertiary and late Mesozoic sands of this region are porous traps for millions of barrels of oil that began as organic matter accumulating in shallow waters of a continental shelf. This shelf was created as the southern shore of the North American continent slowly drifted apart from the southern continent to which it had been joined in Pangaean days. Just as the infant Atlantic Ocean opened up when the Triassic rifts of the East Coast formed, the Gulf of Mexico, initially a narrow embayment of the sea, gradually widened and deepened as the continents drifted apart. The margins of the new continent were left behind by the divergent plate boundary at the center of the opening ocean, left to subside passively and receive sediment for the next 100 million years. The passivity stems from the absence of that plate boundary; the subsidence is from changing patterns of heat flow from the interior of the Earth and therefore called thermal subsidence.

During rifting, the continental crust is thinned as it is pulled apart and extended, so that the heat from the mantle below now flows through a much thinner crust. The temperature gradient of this crust before stretching had an increase of about 30°C per kilometer of depth. After stretching, the rocks of the crust, exceedingly poor conductors of heat, stayed close to the same temperatures, but as a result of thinning, the gradient would have been much steeper, perhaps up to 60°C per kilometer of depth. With time, conductive heat flow would have allowed the crust to relax to an equilibrium thermal gradient. As it cooled, the hot

Sands of sedimentary environments of a passive continental margin.

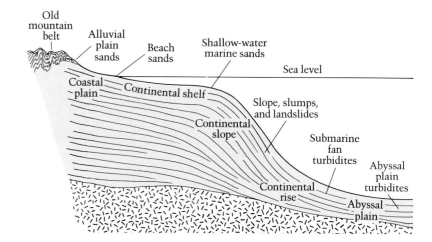

rifted crust contracted. In that contraction lies the initial subsidence, which we can calculate from the coefficient of thermal expansion of rock. The subsidence allowed great thicknesses, of sandstone, shale, and limestone—more than 15 km in some places—to accumulate at the margins of the continent, and the weight of this thick pile of sediment dumped at the continent's edge reinforced thermal subsidence.

The sands of the Gulf Coast were deposited in a complex of shoreline and shallow continental-shelf environments, with turbidites laid down in the deeper continental slope and rise. The composition of the sands reflects their provenance. They are clearly of continental origin and bear a striking resemblance to the sands carried along the lower reaches of the modern Mississippi River. Generations of oil geologists working on the Gulf Coast have shown how the Mississippi has carried the sediments eroded from the watersheds of the Ohio, the Missouri, and all its other tributaries for millions of years. The master stream then drops the sediment at the delta, but longshore currents rework the sand and mud and carry it long distances west of the delta. There it may mix with smaller amounts of locally derived sands from the Sabine and other rivers of Texas and with the major input from the Rio Grande, which drains the southern Rockies. Many of the sand grains are recycled from erosion of older sandstones of the central continent. Others are new to the sedimentary cycle, fresh from an igneous rock of the Appalachians or the

Rockies. The sands from Yellowstone may mix with the sands of Pennsylvania. To tease this information from a rock long buried keeps challenging us, intellectually and economically, to understand more about how the continent grew and to find more oil and gas. How oil and gas are made is the topic of the next chapter, in which we see what happens to sand after it has been deposited and buried by continued sedimentation.

CHAPTER EIGHT

From sand to sandstone

Fashionable brownstone houses in Manhattan are known for their architecture, their history, and their high cost. Although sandstone is far from the strongest building stone, this local brownstone has good color, is easily cut into building blocks, and is available from nearby quarries. Roaming New York City is a continued demonstration to the geologist of the unforeseen uses of sediments once deposited in rift valleys that extended along this locale. The brownstone is an arkose—more than 25 percent feldspar grains—deposited in downfaulted valleys created as Pangaea, the supercontinent, pulled apart to form the proto-Atlantic Ocean in Triassic time, about 225 million years ago.

Sandstone counterparts of New York brownstone are quarried in southern Wisconsin from pure quartz arenites, whose sand grains are more than 95 percent quartz, deposited in shallow seas of the Cambrian period, about 500 million years ago. In eastern Tennessee the building stone is crab orchard stone, a pink or reddish sandstone deposited by rivers of Carboniferous time. Somewhat mysteriously, some of these sandstones do not seem hard or tough enough to be building stone when they are quarried. You can rub individual grains off the surface, and if you are a good hand with a mortar and pestle, you can easily disaggregate a piece into its constituent grains. Other sandstones are everything you would expect from a building stone: dense rocks that resist crushing and clink musically when struck with a hammer.

Regardless of their hardness, all the members of this diverse group of sandstones have one common property: the sand grains are bound together by a cement, a binding material that precipitated around the detrital grains sometime after they were deposited as loose sand grains. Dense, hard sandstones seem to have acquired these qualities long before they were quarried; more friable sandstones earn those characteristics after quarrying. Sitting

Carved ornamentation on a sandstone building, New York City.

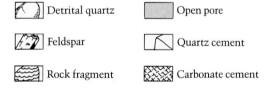

A typical sandstone shows, in this magnified image, more or less rounded grains of quartz, feldspar, and rock fragments, cemented by diagenetic, or postdepositional, precipitates of quartz and calcium carbonate. A small amount of open pore space remains.

out in the rain after being cut, in just a few years—real years, not geologists' millions of years—these light-colored blocks grow a hard, dense veneer that colors them dark brown. Under the microscope the veneer appears as a rind of sand grains cemented with iron oxide. A careful comparison of rind with uncemented sandstone shows that many of the iron-bearing silicates of the sandstone, such as the pyroxenes that were given as examples of rapid weathering in Chapter 1, have disappeared in the rind. Weathering has leached those silicates of their iron and deposited the iron as oxide cement.

Far more important than the occasional use of sandstone for building stone are the reservoirs that sandstone bodies make for oil and gas. Our ability to produce these fluids from the pore spaces between the constituent grains of sandstones depends on a pore space sufficient to hold a large quantity of fluid. There must also be enough flow paths among the pores to allow the oil or gas to flow freely into the drill hole and be pumped to the surface. The porosity, the amount of open space between grains, and the permeability, the ability to transmit flows through pore space, are strongly determined by postdepositional processes. The existence of the oil or gas itself is the result of chemical reactions of organic substances that take place in rocks after they are buried deeply in the crust of the Earth.

The shorthand word for all of these postdepositional processes is "diagenesis," a word coined about a hundred years ago. By that time geologists could see that this varied group of chemical and physical reactions had much to do with the properties of sedimentary rocks and therefore deserved equal ranking with the ordinary genesis of sediments by deposition. From a quiet backwater of sedimentary geology in the early part of this century, the study of diagenesis—mainly powered by demands of the oil industry—has become central to research on the origin of sedimentary rocks. In the good old days, it was worth a brief student presentation in an advanced seminar on odds and ends. That was my introduction, and I was appalled at the paucity of either information or well-worked out theory. One paper set out a "universal law" supposedly governing the order in which cementing minerals were precipitated in all sandstones; the idea was soon debunked. Others offered wonderfully systematic descriptions without benefit of theory. However, we had a few core observations and hypotheses to suggest that this complex but analyzable set of chemical and physical processes called diagenesis could explain much of the geology of burial. More practically, it was becoming obvious that diagenesis was responsible for the low porosity and permeability that made it difficult to produce oil from some "tight," oil-saturated sandstones.

Aside from its practical justification, the study of diagenesis ties sedimentology to a group of emerging concerns about the way the crust of the Earth behaves on continents and continental margins. Diagenesis is most intense in deeply buried sediments laid down in sedimentary basins, regions where the crust has subsided in some linkage with sedimentation. These and other structural movements of the crust require sedimentologists to relate geophysical movements to the chemical and physical changes going on in sedimentary rocks. Because changes in pressure and temperature affect diagenetic reactions, we need to know how these variables behave during burial in a sedimentary basin. Diagenesis, then, brings into play larger aspects of the geology of sandstone: how sands are preserved, entombed, and hardened in the context of the larger-scale movements of continental crust.

PRESERVING SAND

Nature has its own mummifiers, preserving vestiges of former times and events by burying the sediments that record the processes that made them. For example, sedimentation and tectonics work together to preserve river sands. The dynamics of a river system includes erosion of previously deposited sands as well as sedimentation. The definition of an equilibrium profile of a river, its height plotted against distance downstream, requires the river to maintain its downhill gradient, carry all the water supplied to it, and transport all the detritus weathered from its drainage area. The equilibrium river does not accumulate and thus preserve sediment; it keeps its sandbars, channel sands, and floodplain deposits in a constantly changing local configuration that is only as thick as the depth of the channel. In accumulating and preserving sediment, the river departs from equilibrium.

A river in a nonequilibrium state may rapidly steepen its gradient if the region of its headwaters is suddenly uplifted. It will slowly make its gradient shallow in response to a gradual lower-

Top: The longitudinal profile of a river bed, showing the typical fall in elevation from headwaters to mouth. *Bottom*: A river adjusts its longitudinal profile by aggrading, or adding sediment, to its valley when its headwater region is uplifted.

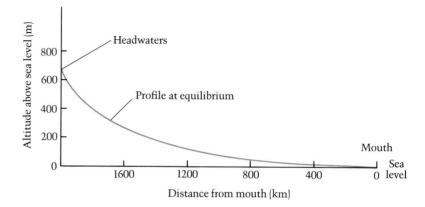

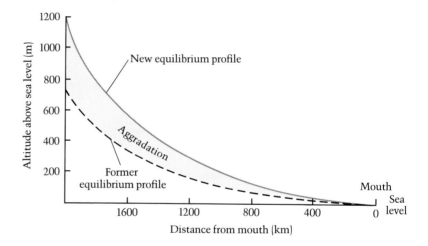

ing of the elevation of its headwaters by erosion. In attempting to recover its equilibrium profile, the steepened-gradient river will deposit more sand, gravel, and silt along its upper to middle course, and the shallowed-gradient river will erode its bed in its upper regions. A stream shows the same behavior after a dam is built: it silts up its upstream reservoir (net sedimentation) and erodes its bed downstream (net erosion).

Sea-level change can accomplish the same steepening or shallowing of the profile by raising or lowering the lower anchor of its gradient. Other things being equal, a rise in sea level will shallow the gradient and a fall will steepen it. The changes in gradient caused by sea-level changes are reflected in deposition or erosion

A river will erode its valley to achieve a new equilibrium profile when sea level drops.

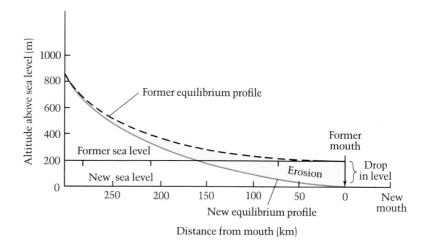

near the river's mouth, depositing when sea level rises and eroding when it falls. Similarly, changes in water or sediment load lead to erosion or sedimentation. Some of these mechanisms operate on a relatively short time scale. In the geologist's long time scale, sea-level changes and tectonic uplifts happen quickly. Weathering is slow, and tectonic subsidence, a sinking of the crust, one of the most important causes of sedimentary accumulation, may be very slow.

Seeing great thicknesses of river sediments suggests that some combination of these factors influenced net sedimentation. In practice, geologists have found that the coupling of tectonic uplift in sourcelands with subsidence below the uplifted area is the major determinant of accumulations many kilometers thick. The uplift of high mountains in the source area of the river creates more rapid and extensive weathering and erosion, which in turn loads the river with detritus. Sedimentary loading from increased river deposition on the Earth's crust causes the middle and lower regions of the river to subside. As the crust subsides, the river dumps more sediment to keep its equilibrium profile.

The positive feedback of these two processes is further enhanced by the flexure, or downbend, of the Earth's crust adjacent to an uplift. Powered by the same deformational forces that created the uplift, this downbend adds to the subsidence initiated by sedimentary loading. The flat river plain of the Ganges hides many kilometers of buried river sediment derived from the ero-

Sandy alluvial fans and plains build up to a new profile of equilibrium after an uplift of the source area.

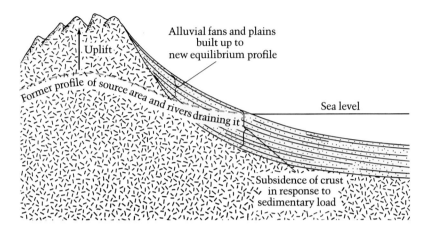

sion of the rising Himalayas in combination with the subsidence of the lowlands to the south of the mountain belt. The Catskill Mountains of eastern New York are made of sandstones deposited in a similar basin 350 million years ago, lying to the west of a high mountain range ancestral to the present Appalachians. The sandstones of these origins are tracers, clues to the uplift and subsidence that allowed them to form and accumulate.

The preservation of beaches or of nearshore marine sands works somewhat differently. The steady state, or equilibrium, of nearshore environments is tied to sea level and subsidence. If subsidence is negligible and sea level is constant, the beaches and bars are in dynamic equilibrium over a time scale of tens of years. During shorter time periods the profile of sediments, in this case, the depth of the sea floor or height of the beach with respect to sea level plotted against distance from the shore, varies in response to storms, winds, and waves.

In a transgression, a rise in sea level that alters a steady-state shoreline, the attack of the waves moves inland and buries the former beach and dune belt. The former shoreline is now under water at a depth that depends on the extent of the sea-level rise and the angle of the submarine slope to the deeper water of the continental shelf. The sands of these offshore environments are covered with new deposits of shallow marine sands until a new profile of equilibrium is reached, with a newly established beach and dune inland of its former position. The first stage of burial of the beach deposits has removed it from direct contact with the environment. The beach sand now sits at some shallow depth in

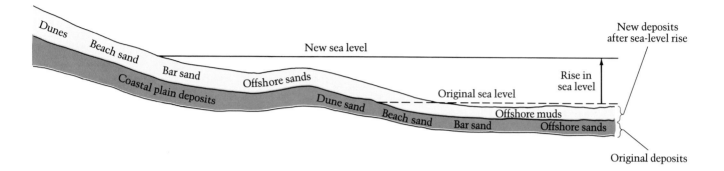

After a rise in sea level, the sea transgresses the former land surface and new deposits of deeper-water environments overlie former deposits laid down at the time of lower sea level.

the sediment pile, separated from the seawater that deposited the overlying offshore sand. In successive events, the beach sand will be buried deeper and deeper as new layers of sediment are laid down.

In a regression of the sea, a withdrawal of seawater from the shallow shelf offshore that results from a drop in sea level, we see different consequences. Now the former beach is high and dry, separated from the shoreline by some distance of newly emerged sea floor, covered with shallow-water deposits. With sea level lower, streams formerly at equilibrium now change their profiles by eroding the coastal plain, including the beach. The survival of the beach deposit depends on the rate and completeness of erosion and the time before the next rise in sea level rescues it. If it remains, it may be covered with dunes, river deposits, or the vegetation of low coastal-plain interstream areas. In any case, in its new environment the remains of any seawater in its pores will be flushed by rainwater infiltering the sediment.

The effects of subsidence are the same for all environments: if sea level does not change and sediment supply is sufficient, the locus of the beach stays the same and the deposit thickens. However, the beach varies appreciably from month to month. It slowly accumulates a gently sloping profile during calm weather and washes much sand out to deeper water during storms. The vagaries of preservation of these variable forms of the beach lead to heterogeneity in a buried beach deposit. Over a longer time scale, short-term variability may sometimes give a deposit showing the storm-wave and tidal-surge bedding characteristic of rough waters and other times preserve the gentler sedimentary structures of lower waves and tides. In the ordinary case, when subsidence combines with sea-level changes, an accumulating section will include sands that mark alternate transgressions and regressions. Some are powered by local sea-level changes that result from tectonic movements; others are attributable to global sea-level rises and falls.

Teasing this information out of a sedimentary sequence gives us both a local and a global history of the shoreline. The sea-level changes of the last million years reflect the waxing and waning of continental glaciers. The low stands, times of lowered sea level, marked by fossil beaches now covered by the sea, correlated with the withdrawal of massive amounts of water from the oceans to be frozen into ice sheets. The high stands, times of raised sea level, marked by shoreline terraces well above present sea level, show the effects of melting ice water running into the ocean. In this way, global climate is deciphered from coastal-plain and off-shore-marine sedimentary sequences.

WHERE THE CRUST SUBSIDES

If you have ever walked a plank over a patch of mud, you have experienced in a primitive way some of the mechanics of subsidence of the Earth's crust. The plank is solid, but it bends under your weight. The mud appears solid, but it flows under the plank as you displace it with your load. In this crude analogy, you are the rock load on the solid but bendable crust, the lithosphere, and the mud is the asthenosphere below, the plastically deformable layer that flows at a much slower than glacial pace. When sediment is loaded on crust, the additional weight is added to the existing load of lithosphere; the asthenosphere, about 100 km below, responds by slowly oozing out from beneath. Because the rock column gradually increases in density with depth, the subsidence does not match the load, and as the load increases, the amount of subsidence attenuates.

Loading by sediment is not usually the only, or even the main, cause of subsidence. Stretching the lithosphere by tension thins the lithosphere and breaks the dominantly brittle upper part of the crust by faulting, creating rift valleys. Lithospheric thinning produces high geothermal gradients, the increase of temperature with depth in the Earth, because the same amount of heat continues to escape the mantle through a thinner crust. The gradient is fixed at the upper end, bounded by the surface temperature of the Earth, which is set by the sun and climate. Hence rift valleys are hot regions of the crust, even hotter if they overlie rising, hot convective currents in the mantle, as they frequently do. We see the surface expressions of this heat in abundant hot springs and volcanoes, and it has also swollen the thinned crust by thermal expansion. We have now set the stage for subsidence.

The initial subsidence of a rift is the downdropped fault basin created by crustal stretching. As the highlands of the rift borders are eroded, thick aprons of alluvial fans, plains, and lakes fill the floor of the valley with sediments several kilometers thick. The sediment loading has begun. Because the life of a rift is short, geologically speaking—rarely more than a few tens of millions of years—the stretching phase decays to stability. As stretching attenuates, so does the thermal expansion. As relaxation sets in, the distended crust cools and thermally contracts. A different sort of subsidence now takes over. The rift valley, floored with land sediments, subsides in response to the crustal contraction and typically sinks below sea level, as the Dead Sea is now. Sooner or later, the sea finds a way in and floods the valley. Marine sands cover what has now become the basin, for it has widened to include the subsiding rift borderlands as well as the valley itself. The shallow spoon-shaped basin underlying southern Illinois and adjacent parts of Indiana and Kentucky exemplifies this kind of evolution as we discussed in Chapter 7 (pages 159 through 161).

The story of the Illinois basin starts about 100 million years before the beginning of Phanerozoic times, the 570 million years of most recent geologic history we call "known" because they contain fossils of shelled organisms. A rift system split the crust in a giant X running through the states of Illinois, Indiana, Kentucky, Missouri, and Arkansas. For as much as 100 million years these rift valleys were actively downdropping along great border faults and receiving enormous tonnages of sand, gravel, and mud distributed as alluvial and lake deposits. After the beginning of the Phanerozoic, cooling and contraction led to subsidence. By Upper Cambrian time, around 500 million years ago, the surface had sunk below sea level, and a broad, shallow sea covered the area. Pure quartz sands were deposited on the broad marine shelf, soon to be replaced by calcium-carbonate deposits of shells. Subsidence gradually diminished, and a relatively thin series of limestones were laid down over the next 200 million years.

Complete quiescence was not to be the fate of the Illinois basin yet. During the Carboniferous period, the long-traveled detritus from the Acadian orogeny, which also provided debris to the Catskill delta at the foot of the mountains, started to reach the midcontinent. As the Acadian Mountains continued to dominate the eastern part of the continent, an enormous alluvial plain extended from Pennsylvania to Oklahoma. Under part of this plain, the Illinois basin, now ever so gently subsiding, resumed filling at a more rapid pace as additional sediment loading progressed. Geological evidence of faults cutting Carboniferous formations indicates that during this long time the original rift faults continued to move at intervals, just as periodic earthquakes in modern times show that the rift faults continue to be restless. The history of the Illinois basin is a lesson in the interplay of subsidence and sedimentation, and we see the result in the preservation and burial of a long-lived sedimentary sequence of sandstones, limestones, and shales that spans 500 million years, a significant fraction of Earth history.

A ride across the flat prairies of southern Illinois brings no hint of this extraordinary history. Although the surface formations exposed along the bottoms of creeks and a few small road cuts tell little about the rocks below, a distant oil derrick reminds the geologist that something valuable lies there. The search for oil and gas has made the map of southern Illinois counties a pincushion speckled with thousands of wells. We can piece some of the story together from the rock samples of those wells and the geophysical logs of the drill holes, even though most of the basin still lies buried, accessible only to the drill and to earthquake waves.

Many other basins have ended their sedimentary history with uplift, deformation, and erosion. The surface-bound geologist wanders among outcrops in the mountains, fitting the jigsaw puzzle of a basin's birth, development, and death to the geologic history of its continent. The rocks of an uplifted sedimentary basin show, at least for one cycle, the geological cycle as James Hutton envisioned it in 1785: an endless series of weathering, sedimentation, burial, deformation, and uplift. Sedimentary basins that have evolved in this way may have only one chance at the cycle of

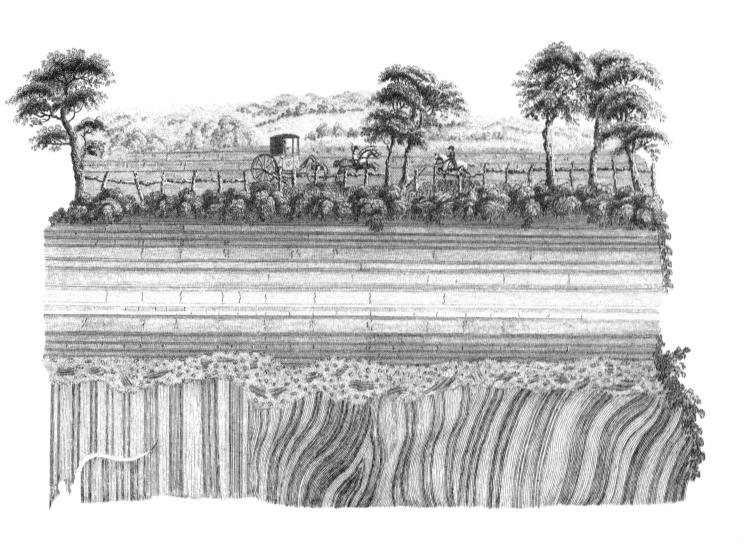

In the late eighteenth century, James Hutton drew this cross section of rocks in Scotland that illustrated his idea of the geological cycle. The rocks below were originally deposited as horizontally bedded sediments. They were then buried, deformed, and uplifted. After erosion planed off the surface, a new geological cycle began with deposition of the upper set of sedimentary layers.

sedimentation, subsidence, and uplift, for once uplifted, they may never again subside. At their demise, the deformation that has crumpled and uplifted them may weld them to a firmly stable terrain of old continental crust that does not get involved in new basin formation—at least so far in the Earth's history. Even with geologic time we must be cautious. Not enough hundreds of million years may yet have elapsed since immobilization of the basin to allow a new episode of continental breakup and orogeny.

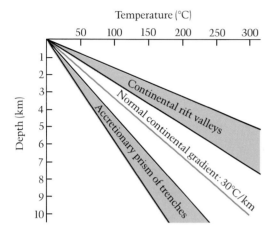

Temperature (°C)

Geothermal gradients of different tectonic settings.

BAKING AND SQUEEZING

As a sandstone sinks in a basin, it moves along the gradient of pressure and temperature in the crust. In a deeply buried formation about 8 km down, a sandstone may be at a temperature of about 250°C and a pressure of about 1800 atm. The baking and squeezing that go on in this pressure cooker severely affect the composition and texture of a sandstone: these changes weld the grains together in tight conformation, with extremely low porosity; for example, clay minerals alter to crystalline micas that begin to resemble the micas of the higher-temperature clan of rocks, the metamorphic schists, gneisses, and quartzites formed at temperatures of 400°C or higher. At lower temperatures and pressures, the changes may be less obvious but are nonetheless pervasive. Some of the diagenetic changes in sandstones seem to start almost at the moment of burial by the next layer of sand, at the normally low temperatures and pressures of the Earth's surface.

A continental crust of normal thickness that is not close to an active plate boundary or a rift valley has an average temperature gradient of 30°C/km depth. This thermal gradient is the result of heat flow from the interior of the Earth generated by the radioactive decay of uranium, thorium, and potassium 40 (a radioactive isotope of potassium that is common in rocks) and by other heat sources such as crystallization of the solid part of the core. The distribution of heat flow in the mantle and crust is governed by the convection currents that slowly overturn the mantle and drive the cool plates at the surface of the globe. At either continent-splitting rifts or those at midocean ridges, rising convection currents bring higher than normal heat to the base of the crust. At subduction zones, where sinking slabs of lithosphere follow downward-moving convection currents, heat flow is low. Normal heat flow, then, is the value for the great proportion of continental surface lying far from the convection currents that drive plate motions.

Geothermal gradients through the crust are determined at the bottom by mantle heat flow and at the top are anchored by the mean annual temperature a few meters below the Earth's surface. Between those points the gradient depends on the thickness and thermal conductivity of the crustal rocks, most of which have

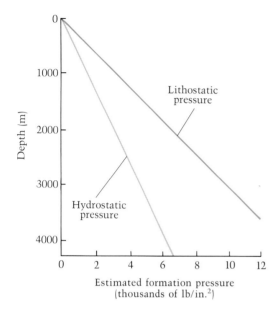

The estimated reservoir pressure versus depth for over 100 wells in a group of Gulf Coast oil fields. Lithostatic pressure is that of the weight of overlying rock, whereas hydrostatic pressure is that of the weight of overlying fluid. Pressures move to lithostatic with increasing depth as pore space is closed.

about the same relatively low value. That low value of thermal conductivity—not dissimilar to that of brick, which takes a long time to heat up but retains its heat for a long time—is responsible for the crust's slow response to changes in heat flow. These facts fit the general assumption that heat is transmitted through the crust primarily by conduction rather than by radiation or convection. In some sedimentary basins, some of the heat may be transmitted by convection—not of the solid rock, as in the mantle, but of pore fluids traveling through volumes of rock.

Estimating the pressure on deeply buried rocks is a matter of calculating the weight of rock and fluid overlying them. For most porous sedimentary rocks in the top 10 km of the crust, the pressure of overlying rock is determined by the density of the rock—most rock-forming minerals are about 2.7 g/cm^3—minus the pore space, most of which is filled with water. The assumption here is that the rock is supporting the overlying rock, whereas the pore water has the pressure of the overlying column of pore water. This implies a connected pore space extending from depth to surface. Although these assumptions may seem overly simple, it is remarkable how well they approximate measured pressures in deep drill holes. Using them, we can then draw two curves for pressure: one lithostatic, with the weight of rock on rock, and the other hydrostatic, with the weight of water on water. For average-porosity sedimentary rocks in the top few kilometers, lithostatic pressure increases at a rate of about 220 atm/km. Hydrostatic pressure is much less, about 100 atm/km on the average.

If the assumption of pore space continuously open to the surface is violated, the pore-fluid pressure starts to approach lithostatic. In many regions of deep burial, shales compact and lose permeability almost completely, shutting off all pore spaces in the rock and acting as a pressure seal for underlying rocks. Thus, although an underlying sandstone maintains its porosity, it will become a closed system of pressure, with the pore fluid at the same pressure as the rock. By comparison with ordinary pore fluid, it is overpressured. Imagine what happens when a drill bites through the sealing shale and penetrates the overpressured sandstone. The overpressured pores now meet the hydrostatic pressure of the column of water in the drill hole: they respond by explosively releasing that pressure. The oil gushers familiar to moviegoers of the 1930s dramatically play out this released overpressure. (My first experience with a pressure release came when I heard a strange sound like a railroad train coming from the drilling well and the driller yelling, "Run like hell!" Then pipe came hurtling out of the hole and all of us were soaked in a mixture of oil, mud, and water.)

FLUIDS, DIAGENESIS, AND FLOW

Children at the beach, some with patience, others with furious energy, explore the sand and discover that if they dig a deep enough hole, water will seep in at the bottom. Everywhere in sands on land is a level below which the pores are saturated with water. At the ocean beach it is seawater; on a river sandbar, the same fresh water as the river itself; on a desert, slightly saline water that lies hundreds of meters down. In marine sands, seawater fills the pores to the surface of the sediment at whatever depth of water it lies. Once buried, the sand grains lie steeped in this medium that had actively transported them to their burial place. The time scale then changes. Although their transport had taken only days, weeks, or a few years, the sands now start to inhabit their burial environment for periods of many thousands or millions of years. Sluggish chemical reactions that barely had a chance to start between sand grain and water during transport now have, if not all the time in the world, orders of magnitude more than their transport time.

The early diagenetic reactions that take place soon after deposition and initial burial may alter a sand's appearance profoundly. At low tide the clam digger leaves the light gray or tan sand of the beach and walks out on the watery, black, tidal flats. The black sediment familiar to residents of so many coastal areas of high-tidal range was long thought to be a muddy deposit rich in black-pigmenting organic matter. I had played in this black stuff on a New England beach as a child—much to my mother's horror—on a long-remembered trip to Boston from Chicago, where Lake Michigan beaches had no tidal flats and no lovely black mud.

I got involved again with this stuff in Barnstable salt marsh on Cape Cod when I visited one of my graduate students there. Robert Berner was spending a summer fellowship at the Woods Hole Oceanographic Institution and had been directed by oceanographer Alfred Redfield to study the sediments in the marsh. What

Digging for clams, Yaquina Bay, Oregon Coast.

Bob found out was a story that unfolded as we walked on the drier, upper parts of the marsh and rowed across the shallow waters of the inlets, digging finally into the flat, black tidal sands. For that is what they were, not muds. They were the same sands that we could see on the low dunes back of the beach that separated the tidal inlet from the open waters of Cape Cod Bay. The blackness was not organic matter—although there was enough of that around—but an amorphous precipitate of ferrous sulfide. A scoop of black sand held to our noses quickly gave the telltale smell of hydrogen sulfide. Berner was to spell out the story of these sands over the next two decades. (Now at Yale, he has become the dean of early diagenesis students and the guru of the present and past members of FOAM, the Friends of Anoxic Mud, recruited from his graduate students and postdoctoral fellows.) As it turned out, the most fascinating complexities of early diagenesis lie in the muds and sands of anoxic, or oxygen-deficient, environments.

The profile, or cross section of sediment, we saw at low tide on Barnstable marsh was many layers of sand deposited by tides. At the top, a thin layer of sand, only a few grains thick, was brown

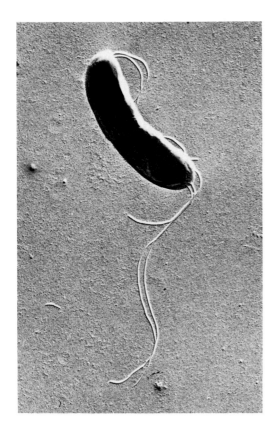

Desulfovibrio desulfuricans, the bacterium that reduces sulfate (SO_4^{2-}) to sulfide (S^{2-}).

with freshly precipitated iron oxide and hydroxide. Immediately below were the black layers, intensely colored by the ferrous sulfide and foul-smelling from hydrogen sulfide. Below the black zone, the sand was light gray, with the occasional glint of a crystal of pyrite, the iron bisulfide of fool's gold. In the Ph.D. thesis that grew from this summer project, Berner teased from his field data and experiments the group of chemical reactions responsible for the succession of differently colored layers.

Just below the sediment surface, the decay of abundant organic matter—the remains of snails, clams, fish, algae, and other denizens of the tidal flats—depletes the dissolved oxygen in the seawater buried in the pores of the sand. Once the oxygen is gone, a new actor enters the scene, *Desulfovibrio desulfuricans*, the bacterium that uses organic matter in the absence of oxygen to reduce the sulfate in seawater to sulfide. The product of this reaction is hydrogen sulfide:

$$SO_4^{2-} + 2CH_2O = H_2S + 2HCO_3^-$$

where we represent organic matter by the formula for formaldehyde. At the same time, iron leached from iron oxide coatings or iron-bearing minerals may be reduced to ferrous form by organic matter:

$$4Fe(OH)_3 + CH_2O + 7CO_2 = 4Fe^{2+} + 8HCO_3^- + 3H_2O$$

where we lump all iron species as ferric hydroxide. Alternatively, iron may be reduced by hydrogen sulfide. Reduced iron and hydrogen sulfide in turn react to precipitate black, amorphous ferrous sulfide:

$$H_2S + Fe^{2+} = FeS$$

and this reaction proceeds as long as there is organic matter to power the sulfate reduction. This sequence explains the black layer. The gray layer below is the result of a different reaction by which ferrous sulfide ages and recrystallizes to form the well-crystallized pyrite by reaction with elemental sulfur, S^0, which is also produced in the black zone:

$$FeS + S^0 = FeS_2$$

The top brownish layer is the result of diffusion of oxygen from the atmosphere above into the top of the reduced zone, oxidizing the iron in the ferrous sulfate to ferric and precipitating ferric oxide and hydroxide.

All these reactions go on simultaneously in the different layers. As sedimentation of sand continues and a sequence accumulates, the brownish oxidized layer stays at the surface while an individual sand grain, originally an oxidized grain, sinks through the black zone and then into the gray zone, below which it remains stable and unchanging. The geologist coming along millions of years later has to infer this complex set of interactions from one relic only, the little cubes and crystalline aggregates of pyrite.

The early diagenesis of organic matter and sulfur to form pyrite is only one of many transformations in a young, barely buried sand. Another is the formation of calcium carbonate or silica cement. Many marine sandstones contain several forms and types of shell fragments, including relatively unstable forms of calcium carbonate: forms with significant amounts of magnesium substituting for calcium in the crystal structure; fragments constructed of aragonite, the metastable form of the substance; and most commonly, forms with organic-tissue remains within the meshwork of microcrystalline layers secreted by the organism. All these are unstable relative to the pure calcite mineral that is the stable form. In early diagenesis, the unstable forms may dissolve and reprecipitate as clear, large calcite crystals filling the pores and thus binding the grains together.

In some sands the shells are of diatoms or radiolaria, unicellular organisms made of a common variety of biochemically secreted opaline silica. These shells may dissolve in the early diagenetic pore waters and reprecipitate as quartz: normally, this quartz is precipitated on single quartz crystals, the commonest kind of sand grain. When deposited as an addition to the existing crystal structure, it acts as if the original grain had continued to grow; the overgrowth of new quartz may not be distinguishable under ordinary or polarized light but may betray its presence by the sharp, planar crystal faces it characteristically assumes. These kinds of clean, planar crystal faces must be diagenetic, for crystal faces of detrital grains are always worn and rounded by abrasion during transport. The sparkling appearance of some sandstones is attributable to these kinds of overgrowths.

The chain of early diagenetic reactions by which pyrite is precipitated.

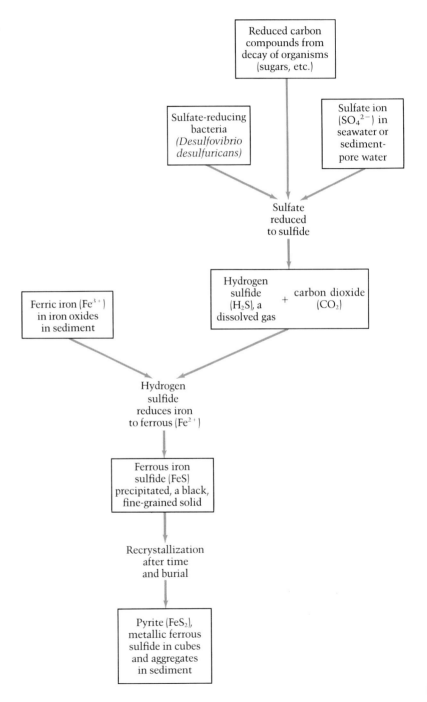

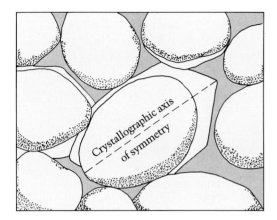

 Open pore

Detrital grain

Diagenetic quartz overgrowth

Quartz sand grains are enlarged by diagenetic quartz precipitates that are in crystallographic continuity with the original grain. Such enlargement may create the familiar sharp crystal faces of quartz.

Our best means of identifying quartz overgrowths is by cathodoluminescence, the light emitted by some substances when bombarded by electrons. Because the luminescence is sensitive to small traces of impurities, we can use it to distinguish the luminescent detrital grains that originated by crystallization at high temperatures (at which impurities are incorporated) from the nonluminescent overgrowths formed at low temperatures (containing virtually no impurities).

Calcite cement and quartz overgrowths are found together in many sandstones as redistributions of material already present in the sediment. If a sand is deposited in a carbonate-producing environment, calcium carbonate may be precipitated around the grains, filling the pore space and cementing the sand to a hard sandstone in a short time. In sands that are deposited in hypersaline areas where seawater is evaporating, some of the minerals precipitating as sea salts will fill pore spaces and cement the sand. Varieties of calcium sulfate, with different crystal structures, anhydrite ($CaSO_4$) or gypsum ($CaSO_4 \cdot 2H_2O$), are the most common evaporite cements. The catalogue of early cements is long and includes iron oxides, calcium and iron phosphates, iron silicates, and a host of rare minerals.

Sands can lose constituents as well as gain them. Feldspar, which is commonly a minor constituent of quartzose sands but an important fraction of arkosic sands, may be unstable in the pore waters of early diagenesis, especially in the alluvial sands that are exposed between floods to infiltration by rainwaters. Rainwater will weather feldspar sand grains in the sediment just as it weathers granites in an erosional environment, as we saw in Chapter 2. In extreme cases, such as rivers running through tropical rain forests, the sands temporarily deposited on sand bars and floodplains may lose almost all their feldspar grains, changing from an arkose to a pure quartz sand. In these ways, early diagenesis can play havoc with the researcher trying to figure out a sand's provenance. In similar sands in arid regions, the weathering of iron silicate grains and the reprecipitation of iron oxide coating the grains produces the red and brown sandstones beloved of Hollywood directors doing westerns.

We can see the importance of fluid flow through pore space in this recital of the minerals that are added and subtracted from sands early in their diagenetic history. Moving fluids do much of

Sunrise at Alabama Hills with the Sierra Nevada Mountains in the background. Iron oxides have colored the sand derived from weathering.

the chemical work of diagenesis. In sands deposited on the continents, the fluid is meteoric water, water that originates in the atmosphere as rainwater, infilters the soil, then travels underground, sometimes for great distances. We see a striking example of such groundwater movement in the Dakota sandstone of the western high plains. This Cretaceous sandstone outcrops in the upturned slabs of rock that appear in the foothills just below the heights of the Front Range of the Rockies. The Garden of the Gods near Colorado Springs is a beautiful example of this kind of outcrop of formations that have been folded up to a position where the bedding is now vertical. Rainwater enters the formation at its outcrop and travels down the dip of the formation. Here, under the pressure of its relatively high elevation, compared with its elevation buried under the Great Plains, the meteoric water moves hundreds of kilometers to the east, its progress punctuated by the numerous water wells drilled to supply the needs of the people of the Dakotas, Nebraska, and other plains states.

As the groundwater moves, it mediates chemical change in its host sandstone. As they infilter near the surface, the meteoric

In the Garden of the Gods, Colorado, upturned beds of sandstone serve as a route for surface waters to infilter the formation.

waters weather and dissolve the small amount of feldspar originally deposited. As the water moves down, it begins to warm along the normal continental geothermal gradient under the high plains. At its deepest, the water may reach a temperature of about 100°C and a hydrostatic pressure of over 800 atm. Water in these circumstances reacts relatively quickly with the minerals and may dissolve significant quantities of quartz, which makes up much of the sandstone. As the water migrates, still under its pressure drive from the outcrop, it follows the structure of the shallow basin and moves up a low slope to the east, now cooling as it goes. The water that was saturated with dissolved silica at 100°C is now supersaturated at 70°C and precipitates the excess as quartz overgrowths—in some places these are added to quartz overgrowths that may have formed during early diagenesis. The resulting sandstone has lost a large fraction of its original porosity and becomes a hard, tightly cemented rock.

Like most familiar substances, quartz becomes more soluble as temperature increases. Carbonate, on the other hand, acts in what we might call a counterintuitive manner: it's solubility decreases as temperature increases. Thus, in the Dakota sandstone, calcium

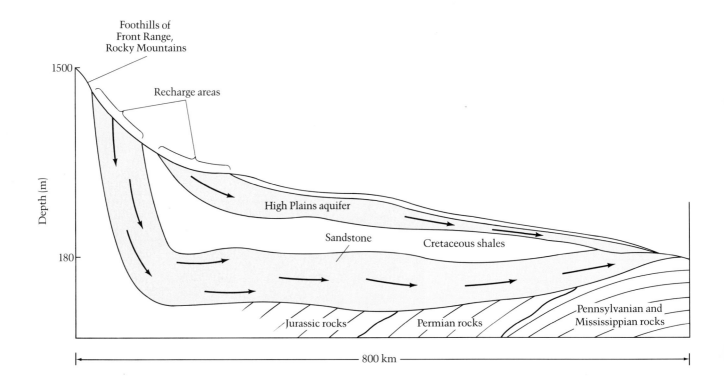

Groundwaters move through deeply buried sandstones under the western high plains of the United States, entering (recharging) in the foothills of the Rockies and discharging by springs and seepage to the east.

carbonate precipitates as the water reaches the deepest, hottest depths and dissolves as it moves upward and cools. The complementary dissolution behavior of these two common minerals explains how, in many situations, diagenetic quartz appears to replace diagenetic carbonate, and vice versa. The replacement is obvious to the petrologist who peers down a microscope at the diagenetically altered grains. The replaced grain appears corroded and embayed by the replacing mineral, which may show sharply defined crystal faces.

In the later stages of diagenesis, which we see in the Dakota sandstone under the Great Plains, fluid movements originate in convection currents driven by heat sources in the lower crust. Under normal geothermal gradients, the heat difference between the bottom and top of a sedimentary basin is not great enough to drive the kind of convection we are familiar with, where the lighter, warmer fluid rises and the heavier, cooler one sinks. Although the minimum temperature difference may exist, a con-

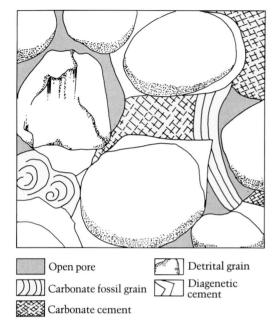

Open pore	Detrital grain
Carbonate fossil grain	Diagenetic cement
Carbonate cement	

A sandstone with calcium carbonate fossil shells may show evidence of a sequence of chemical events during diagenesis. Carbonate fossil grains dissolved as quartz cement was added to quartz grains, in part replacing shell material with the quartz. Carbonate cement is also replaced by quartz, indicating that the carbonate cement was precipitated at an earlier stage than the quartz cement.

vective flow requires also a minimum ability for the medium—in this case porous rock—to transmit the fluid. The cemented, compacted rocks lose their porosity, which significantly reduces transmissibility, or permeability, in most basins.

Where geothermal gradients are higher than normal and coupled with sufficient permeability to transmit flows, convective overturn, the rising of warmer, less dense fluid combined with the sinking of cold, denser fluid, may transport large quantities of fluid through sandstones. The typical heat source is an igneous intrusion at depth. The still-molten or recently congealed igneous rock heats the sedimentary section into which it has intruded, warming the pore fluids in the adjacent sedimentary rocks. Convective heat and fluid transfer may also be the result of crustal stretching, which can raise geothermal gradients. Both of these heat sources are associated with rift valley basins, which are burial environments prone to convective activity. We can trace the diagenetically formed feldspars that commonly occur in the arkosic sandstones of rift basins to these rising hot waters.

The newest rage in studies of subsurface water moving through rocks is the squeezing of water from thick sedimentary sections by the sudden addition of weight from great slabs of rock superimposed on those sections by overthrust faults. Long familiar to students of the Alps and other mountain belts, overthrust sheets are a sign of the regional large, horizontal compressive stresses associated with mountain building. As the crust is compressed, it breaks along low-angle, almost subhorizontal, faults and shoves the upper sheet many tens of kilometers over the lower, sometimes doubling the thickness and, over a short geologic time, loading the underlying rocks with a huge new weight. We can see this weight as another variety of overpressuring. Although the overpressuring may be confined temporarily by the relatively low permeability of the rocks through which the fluid has to flow, it gradually dissipates as the fluid moves slowly out, either laterally to the lower-pressure regions in front of the overthrust belt or vertically through the overthrust sheet itself. Calculations show that enormous quantities of fluid may be expressed from the lower section by this tectonically caused increase in pressure, which results also in the transport of fluid from warmer to cooler regions. This mechanism is of more than passing interest to mineral-resource geologists because it may account for a number of

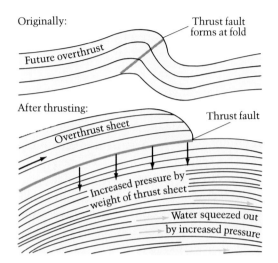

Fluids may be pressed out of a region of rock beneath an overthrust fault as the sudden imposition of additional weight compresses the rock and decreases pore space.

metal-ore deposits found in sedimentary rocks. The lead, zinc, copper, and sulfide that make up these deposits are thought to have been dissolved from the deeper section by the hot waters, transported laterally to adjacent cool sedimentary rock regions, and there precipitated in porous sediments.

PRESSURE SOLUTION

As early as the late eighteenth century, James Hutton and other pioneering geologists described a process of intense dissolution at grain contacts under lithostatic pressure. They saw grains abutting each other with various kinds of interpenetrative contacts, none of which could have formed as a result of sedimentation from a current. These interpenetrative contacts were along plane or concave–convex areas, whereas the grains of a freshly deposited sand are in contact with their neighbors along narrow tangential interfaces or points. Long before enunciations of thermodynamic theory in the nineteenth century and the growth of chemical thermodynamics in the early twentieth century, geologists had concluded that the stress of lithostatic pressure on the point contacts of grains affected the solubility of the substance so that it preferentially dissolved at the points of greatest stress.

This effect on solubility remained a mild curiosity, buried in the unread literature, until 1956, when Milton Heald of West Virginia State University realized the potential importance of pressure at grain contacts for the cementation of sandstones. With an interest in both structural geology and sedimentology, Heald had been looking at some quartz-rich sandstones of the structurally folded Appalachian Mountains and trying to determine how some of them were so completely cemented by quartz. Everywhere he saw the evidence of pressure solution welding the grains. To measure the extent of the process, he devised a parameter, the minus-cement porosity. First, he measured the volume occupied by pore-filling diagenetic cements that had been added to the sand, then he added this volume to the present volume of pores. This total volume was the total pore space remaining between detrital grains. He found that sands that would have been deposited with porosities of 35 to 40 percent had been reduced by interpenetration of grains to porosities of less than 10 percent. Heald's work caught on. More and more geologists, especially those work-

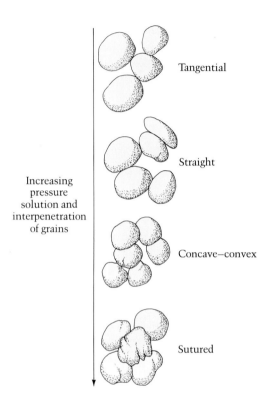

Tangential

Straight

Increasing
pressure
solution and
interpenetration
of grains

Concave–convex

Sutured

Types of grain contact that result from pressure
solution.

ing for oil companies, turned to this explanation for sandstone
cementation.

One oil company researcher, Peter Weyl, held a Ph.D. in physics
and went at the problem theoretically. Hired by Shell Research in
Houston to work with geologists on a range of problems, Weyl was
piqued by the mechanism of pressure solution. Just a few years after
Heald, Weyl proposed a theory in which he related stress, preferen-
tial solution, and the kinetics of dissolution to time and burial. He
zeroed in on the necessity for transport of the dissolved material
away from the grain contact. How could transport occur in what
appeared to be an exceedingly thin film—of the order of a few tens or
hundreds of nanometers—that allowed only limited slow diffusion?
Even long geological times would not be enough to produce some of
the textures the microscopists had seen. Weyl seized on an observation made by Heald and some of his students. They had noted the
much greater intensity of pressure solution in sandstones with small
amounts of clay between the quartz grains. Weyl saw how thin films
of clay and water at the contacts could provide multiple channels for
diffusion and speed up the whole mechanism to reasonable geologic
rates.

This marriage of theory and geological observation set the tone
for an ever-increasing flow of papers on the thermodynamic basis
for pressure solution, the experimental testing of various mechanisms at different pressures and temperatures, and more observations of the geometry and geological distribution of pressure solution. Electron microscopes and cathodoluminescence have been
used to study the contact regions more closely. The importance of
pressure solution has extended beyond the diagenesis of sandstones and limestones—for they too show abundant evidence of
pressure solution—to mechanisms of rock deformation. Structural geologists are sure that this mechanism is partly responsible
for the ways in which rocks under high pressure and temperature
bend and recrystallize.

The other half of the pressure-solution story is the fate of the
material dissolved at the contacts. Once the dissolved silica gets
out of the contact region, it is in open pore space: under normal
hydrostatic pressure. If the pore space were originally saturated,
that is, at chemical equilibrium with respect to quartz, as analyses of most pore waters show, then the new solute provided by
pressure solution diffusing in from grain contacts would super-

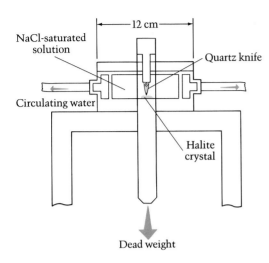

Diagram of knife-edge experimental pressure-solution apparatus.

saturate the solution, that is, give it more dissolved silica than at equilibrium and thus cause it to precipitate the excess. If this input continued, eventually quartz would precipitate, tending to move the pore waters back to equilibrium, probably as overgrowths. About the time when Heald and Weyl were exploring the phenomenon, I had been working on the diagenesis of sandstones in the coal basins of the eastern United States and noting the degree of pressure solution. A complementarity seemed clear: where pressure solution was intense, the sandstone became cemented by the welding of detrital grains along grain contacts; in adjacent regions, the quartz derived ultimately from the pressure solution precipitated at overgrowths. This complementarity has now been shown for a good many sandstones and explains satisfactorily the loss in porosity of sandstones by the combination of dissolution and reprecipitation.

Just a few years ago, I was able to satisfy a long-held wish to do a different sort of experiment on pressure solution. The standard experiment was to load a cylinder of loose quartz sand in some kind of hydraulic jack in which both the directed pressure, along one axis, and the confining, or hydrostatic pressure, could be varied. Quartz is slow to dissolve, and natural sand grains or crystal fragments of quartz are poorly standardized materials to work with. Ryuji Tada, a postdoctoral fellow from the University of Tokyo whom I had first met in Japan, came to Cambridge to work out with me a new way of tackling the problem. Our choice was to build a knife-edge apparatus in which the edge would impinge on a flat plane of a crystal face of a substance whose solid and surface properties were well known. This was the geometry for which, years before, the strain distribution in an isotropic solid had been completely worked out. The material was to be a single crystal of halite, NaCl, one of the best-characterized solids of chemists. The results were startling.

We did produce a dissolution groove in the halite under the knife edge, but its character was surprising. It was a double groove formed along each side of the edge with a slightly raised ridge directly under the center of the knife. Electron microscopy of the groove and ridge revealed etching and irregularities of the surface produced by solid deformation under the knife and dissolution along the free surfaces adjacent to the edge. We could now see that the pressure was straining the solid, making it more soluble,

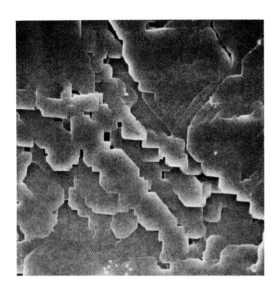

Scanning electron micrograph of the ridge below the knife edge of the pressure-solution apparatus; microcracks dissolve along planes determined by solid deformation.

but that the main scene of dissolution was not at the main contact surface but at its edges, along the free surface where the deformed solid extruded from beneath the knife. Under the knife itself, the solid was hardly affected by dissolution—only by solid deformation. These results made it easy to explain how the extremely slow rate of diffusion along a thin film at the contact could accomplish so much transport. Our final step was to extrapolate to quartz spheres, where we deduced brittle granulation at grain contacts that accompanied free surface dissolution at the edges of the contacts. It is always pleasing to explain a natural process, so complex in detail, by a relatively simple model based on theory and experiment—both suggested by observation.

There is much more to the diagenesis story. After they are deposited, sandstones may go through many different geological histories. To reconstruct these, we are progressing to large computer models for the diagenesis of sandstones in various sedimentary basins. The large databanks of these models carry all the fundamental thermodynamic properties of minerals possible in a system of multiple chemical components. These data are then used to compute changes in temperature and pressure along specific burial paths. Even more complex is the coupled modeling of chemical changes linked to the flow of pore waters; here, the rates of a multitude of chemical reactions are tracked relative to flow rates. Geologists and geochemists are emerging as specialists in computer modeling, devoting most of their waking—and perhaps dreaming—hours to devising better programs on the one hand and hunting more accurate data on the other. These programs take significant time to run, even on supercomputers. The latest wrinkle is the translation of some programs from mainframe to personal microcomputers. (All you have to do is set the program running on your PC before you go to sleep and hope that it is finished by breakfast time.) No matter how well the program runs, however, its utility for prediction of the real world depends on the accuracy of the fundamental chemical and geological data.

The practical test of utility is faced most sharply by the geologists working in the petroleum industry. With millions of dollars at stake they frequently have to recommend on the basis of predictive programs where to drill a sandstone that is deeply buried and inaccessible to ordinary mapping. Sandstone geology and oil findings are intimately related, as we see in the next chapter.

CHAPTER NINE

Oil, gas, and tar sands

I n the first third of this century the world's love affair with oil began. After World War II, the love affair turned into an obsession as every continent was ransacked for new oil fields. Throughout this period of furious exploration, sandstone was the major target for several reasons. Few geologists or wildcatters in the early days thought that limestones could hold oil.

Limestones, rocks formed of calcium carbonate, largely the remains of shelled organisms, seemed to drillers to be dense, low-porosity rocks inhospitable to oil. We now know better, and porous limestones are as important as targets for the oil industry as sandstones. Abundant gas sands were also neglected until pipeline technology evolved, and tar sands, which contain huge reserves of asphaltic hydrocarbons that will not flow without artificial heat, still frustrate our efforts to exploit them efficiently.

Sandstone alone became the object of all searches. Some sandstones flowed dark, heavy oils. Others explosively gushed lighter oils mixed with gas. It is no surprise that the study of sandstones has been inextricably intertwined with the search of oil in this century. Most of the world's sandstone sedimentologists work for oil companies. Most academics who work on sandstones have had their research supported at some time by oil companies. Since 1920 a major forum for presenting research on sandstones has been the American Association of Petroleum Geologists. The unwieldy name of the major organization of American sedimentologists, the Society of Economic Paleontologists and Mineralogists, reflects its history as a subsidiary of the petroleum geologists' association for geologists with specialized interests. The petroleum industry, not only employs many thousands of geologists but, when times are good for the industry, lures many undergraduates to major in geology.

Natural oil seep near Filmore, California

The petroleum geologist wants to know first if a sandstone contains oil or gas and second if it is sufficiently porous and permeable to allow the oil or gas to flow out into the well. The first question entails a series of approaches to the origin of oil. How and when does it appear in a sandstone? Organic geochemists and geologists have worked together to map the route from organic matter to oil and gas. Plant and animal remains are deposited with the original sediment and are then transformed in a series of geochemical reactions, first in early diagenesis and then in later diagenetic stages of higher temperature and pressure.

The question of porosity and permeability leads us directly to the many ways in which diagenesis, superimposed on original textures, fills or reduces pore space and blocks openings from one pore to another. The petroleum industry long ago learned to alter by engineering such properties as permeability, thus rectifying nature's "mistakes" in creating oil fields with insufficient flow rates for pumping. Reservoir engineers have devised a host of measures to enhance permeability and induce the viscous oil to flow into the well. They may artificially fracture the buried rock or pump acid into the formation to create new flow channels. To do so, they have to know much about the physical and chemical properties of the sandstone.

SOURCE ROCKS AND OIL FORMATION

The odd thing is, sands contain relatively little organic matter when they are newly deposited; even the black, sulfidic, tidal-flat sands do not contain very large amounts of organic matter. The place to go for organic matter is mud, and the best muds come from sedimentary environments low in oxygen. There, the bits and pieces of life accumulate as dead cells, excreta, and finely divided particles of what once were the stuff of organisms, proteins, amino acids, fatty acids, and chlorophyll. Little of this organic material remains intact, for the bacteria, whether aerobic or anaerobic, have broken down everything they can. After the early diagenetic stage, bacterial activity slows to an imperceptible pace. The remains of organic matter include small building blocks of life, the amino acid residues of proteins, the remains of fatty acids, and other labile, or unstable, organic molecules—some hydrocarbons among them. In addition to these remains is a

Natural oil seep, now oxidized to tar, at Goleta Beach, California.

more refractory, decay-resistant group that includes some resins and the remains of cutin, chitin, and other collaginous materials in the shells and skins of organisms that help to protect them from attack.

The mud, loaded with these fractions of organic matter, starts its slide down the geothermal gradient and at the same time gets compacted by the weight of overlying sediment. During the compaction, water with solubilized organic matter is expressed from the mud into an adjacent, more porous sediment, such as an overlying sand. The mud, about to become shale—the sedimentary rock formed by the compaction, water-expulsion, and hardening of mud—is the source bed of the oil that inevitably migrates elsewhere. Much of the proto-oil that escapes the mud works its way to the surface—it is, after all, a low-density material that floats on water—and seeps out to be oxidized or swept away. Oil seeps are commonplace all over the world. In East Asia, oil seeps were used for lamps and lubrication more than a thousand years ago, long before the "first" discovery of oil by drilling in 1859 in Titusville, Pennsylvania. The beaches of Santa Barbara, California, had tarry residues from seeps long before the ill-fated blowout of an offshore drilling rig in the 1960s.

These elemental facts of oil occurrence that became well known by 1900 were encompassed in the beginnings of a general theory of oil formation, which included three geological requirements: source sediments containing organic matter, migration pathways by which oil could move from source to permeable rocks, and the reservoir, a formation of permeable rock that would hold the migrated oil. Source, migration, and reservoir remain the trinity of oil finding and are rarely separated. However, the first question we might ask of an area to be explored is: Are there any source beds within striking distance in this basin? If organic matter is the key, we should look for black shales containing appreciable quantities of organic matter, normally 3 to 4 percent. Most of the dark organic matter remaining in these shales descends from the refractory fraction of the original assemblage, the materials resistant to decay or transformation. The labile, or unstable, hydrocarbon fraction has long since gone; it has either migrated or it has been decomposed by bacteria. In some organic-rich shales, however, enough easily combustible material remains that the rock will burn. The oil shales of Colorado and Wyoming are examples of these kinds of organic-rich rock, so rich in hydrocarbons that they release large quantities of oil when heated, thus constituting a large oil resource for future exploitation.

No evidence suggests that any sands are deposited with so much organic matter that they could be considered source beds. This may be partly because the type of organic matter preserved in tidal-flat and alluvial–deltaic sands yields little hydrocarbon after early diagenesis. The organic matter represented by degradation of land vegetation normally moves along a chemical route in diagenesis that takes it to peat and coal rather than oil and gas. These carbonaceous materials contain a large fraction of oxygen-containing carbon compounds that are descended from the cellulose and lignin of plants. For most oils, the source organic matter is instead algae: its remains contain more of the lipid, or fatty, kind of organic compounds. These materials tend to form relatively pure hydrocarbons when they alter during diagenesis. Thus, the typical association of organic matter that accumulated in backwater swamps of rivers with channel and floodplain sands yields coaly layers. Because freshwater algae may be numerous in the standing water of such swampy areas, a small fraction of oily matter may also be found.

Another reason why sands are not good source beds is their high permeability. Whereas a mud quickly seals in its contents as

Cross section of a coastal plain and nearshore marine area. Arrows show downward migration of ground water from landward formations to seaward ones and the transition to marine-pore waters.

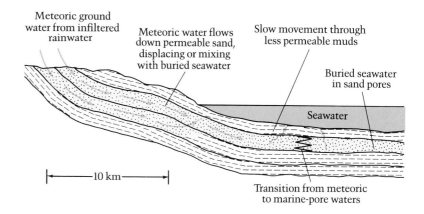

it compacts and loses permeability, a sand loses them. As we saw in Chapter 8, any fluid in a continental sand is quickly transported by meteoric waters, the rainfall that infilters the ground. Marine sands are not penetrated by meteoric waters except along the shore, where meteoric waters from landward formations may invade marine sands for distances ranging from several to tens of kilometers. These fresh waters are able to displace the original seawater buried with marine sands as they move from their entry by infiltration into the outcrop belts on land of sands that slope downward to and are permeably connected with the marine sand still undersea. Marine sands far from shore are rarely invaded in this way but remain connected by permeable channels to the overlying seawater early in their postdepositional history. We are left with an interesting conclusion: sandstone is an excellent reservoir for oil that has formed somewhere else, most likely in a shale. We now have another problem. How did the oil get to the sandstone?

THE JOURNEY FROM SHALE TO SANDSTONE

Thanks to such instruments as the coupled mass spectrometer–gas chromatograph, we can now fingerprint organic compounds and thereby trace the genealogy of an oil. Extremely slight variations in the ratios of different hydrocarbons and their associated

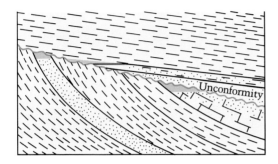

Cross section of oil traps formed along an unconformity surface as oil migrated upward in the formation below the unconformity and were trapped by impermeable beds laid down after the period of erosion represented by the unconformity.

trace metals allow the matching of sandstone reservoir oils and their source beds, which still have enough oil residue left to compare with the reservoir fluid. When we are able to match source bed and reservoir geochemically, we see that sources normally occur not only in the same basin and no farther apart than a kilometer vertically or tens of kilometers laterally, but also in the same general stratigraphic sequence, a succession of strata deposited under broadly the same geologic conditions over a given time interval. In some oil fields, source and reservoir are brought into geological conjunction by faulting or by erosional unconformities, which are gaps in the rock strata, places where one stratigraphic sequence is succeeded by another. Faults may transport the source bed laterally or vertically. Unconformities, produced by the erosion of an underlying series of rocks, followed by sedimentation of a overlying, younger series, may remove large packets of rock intervening between source and reservoir.

These same deformational events serve to create migration paths from source to reservoir. Oil may leave the source via the fault and move up along it to a permeable sandstone that abuts the fault at a higher position. Not all faults are conduits, however; in some oil fields faults serve as seals that prevent the oil from leaving a reservoir. Determining whether a fault is a conduit or a seal is difficult and usually appears evident only after we have made the link between source and reservoir on other grounds. Unconformities may strip away beds above the source and expose it to erosion, thus allowing the oil to escape or oxidize. Petroleum geologists search for unconformities that have not permitted this oxidation but are buried and form a connection between source and overlying reservoir.

No special geological events are needed to provide migration routes for oil. During the life of a sedimentary basin, the lateral and vertical geometries of sedimentation provide enough connections among the different rock formations, both permeable and impermeable. Like faults and folds, unconformities provide migration paths and occur commonly with both sedimentation and tectonic deformation. Like any complex interaction, however, the precise migration path of a specific oil is difficult, if not impossible, to describe. In practice, most oil geologists worry more about the behavior of oil and gas once they arrive in the reservoir than about their journey there.

Idealized structure maps *(above)* and sections *(below)* of several kinds of anticlinal folds. Oil and gas may be trapped at the tops of such folds by impermeable beds overlying the permeable beds of the traps. Different degrees of folding are shown, from a gentle crenulation *(left)* to asymmetric steep folding *(right)*.

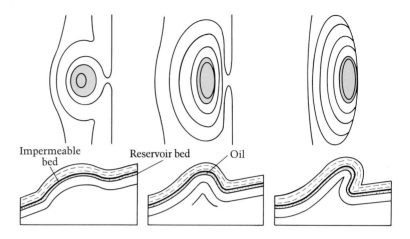

Predicting the presence of a reservoir and its characteristics is the key to success in drilling. Specifically, exploring for oil means searching for a trap. Oil traps are arrangements of strata that hold the fluid and prevent it from working its way to the surface. Not long after the drilling of the Drake well in Titusville, Pennsylvania, in 1859, geologists began using the structural theory of oil traps as their guide: analysis of numerous oil fields had shown that the sandstone reservoir was capped by an impermeable shale barrier and that both formations were folded into an anticline, a convex upward fold of the formations; oil and gas migrate upward to the top of the anticline and are there immobilized in the trap. Remarkably, this simple rule ("find the anticline") proved to be extraordinarily successful, so much so that it was still the mainstay among conservative oil geologists in the middle of the twentieth century.

Although structural theory has not been supplanted—there are too many thousands of oil fields located on anticlines for that—it has been joined by other models of reservoir formation. Some of the famous discoveries in Texas and Oklahoma proved to be accidental stumblings on stratigraphic traps, not simple anticlines. Some of these traps were formed by a sandstone bed pinching out, that is, laterally thinning to the point of disappearance in a body of shale. Others resulted from a loss of permeability caused by extensive cementation of the sandstone. Geologists started to

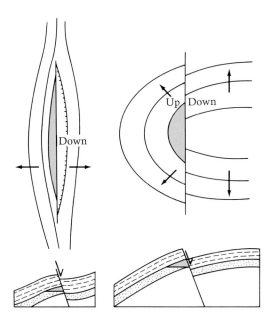

Oil traps formed by faults cutting a low fold *(left)* and a steeper one *(right)*.

recognize other possibilities such as traps formed by faults and unconformities, and by huge salt domes squeezing upward from a deep salt formation and punching through overlying sedimentary rocks.

Wildcatters, oil explorers who drilled into new territory—frequently on the slimmest of chances—made important contributions. Many were not bound by any geological theory at all; instead they followed hunches based on surface characteristics, feelings in the pit of the stomach, or forked sticks. Their frequent successes masked their more common failures but also demonstrated the ordinariness of finding oil in sedimentary basins. Retrospective analyses have shown that purely random drilling in the United States might have resulted in much the same pattern of discoveries as actually happened with the combination of pure geology, geology plus hunch, and pure hunch. That result vanishes, however, when we consider modern exploration for very large oil fields in places such as Alaska's North Slope, the Saudi Arabian desert, or the offshore continental shelf. No one but industry giants—with resources larger than those of a great many countries—could attempt such exploration, where the costs of drill holes and drilling platforms may exceed $100 million. Today, the wildcatters who are left in small oil fields—where drilling holes may not cost much more than a million dollars—are sophisticated operators. Some are geologists themselves; others know a great deal from experience or hire a consulting geologist. Finding a high-quality oil is another goal.

MATURATION, OIL WINDOWS, AND GAS FIELDS

Differences in crude oil quality and the ratio of oil to gas both depend on the temperatures and pressures of oil formation and migration, factors we lump under the term *maturation*. Nature's cracking—breaking more complex organic compounds into simpler ones—is the slow result of heating along the geothermal gradient. When refinery engineers discovered catalytic cracking of crude oil hydrocarbons, they were simulating nature but speeding up the process. They could now efficiently and rapidly produce the short-chain hydrocarbons from octane to methane that were needed for internal combustion engines.

Stratigraphic traps formed by permeability differences. Maps *(above)* and sections *(below)* show a lens of permeable rock surrounded by impermeable rock *(left)* and a pinchout of permeability in a dipping bed *(right).*

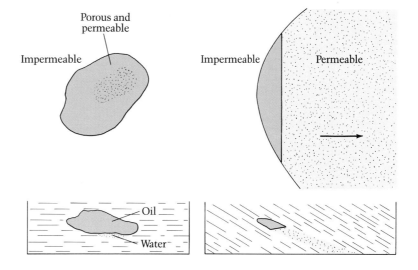

In nature, heavy, that is, denser, viscous crudes had not been cracked as much as the more mature, less dense, less viscous, and more volatile crudes that came from higher temperature formations where cracking had gone on at higher temperatures and for longer times. This observation led to a maturation concept that envisioned light crudes as the product of longer burial times at greater depths and at higher temperatures. As the exploring drills bit deeper into sedimentary basins, the oils would become lighter and richer in gas. Shallow, recently deposited formations would contain heavier, immature oils. But some of the more mature crudes proved to be no older than less mature crudes. Some light crudes were accompanied by large quantities of gas under high pressure and other fields were virtually all gas, with only small amounts of oil. The light crudes might also be associated with formations that were now at lower than expected temperatures. Could they have been cracked by other means or were they once hotter?

By the mid-twentieth century, when the maturation concept was being widely applied to the search for oil, some insight was available from another direction. Coal geologists had long struggled with the geological significance of coal rank, a measure of volatile content that classifies coals from the low-rank brown

Flow chart of the processes of oil formation and maturation that are correlated with stages of burial of source and reservoir rocks.

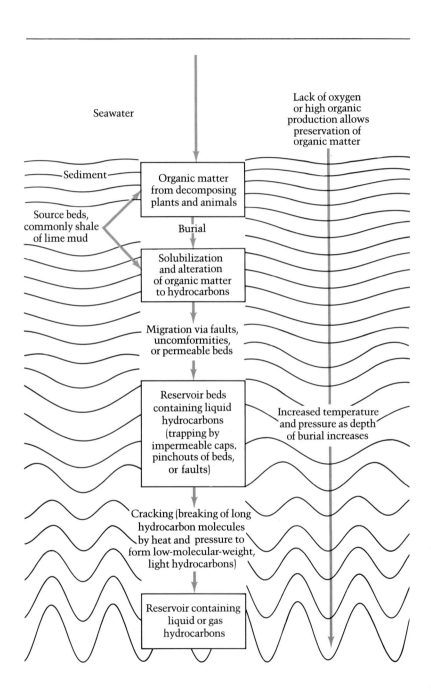

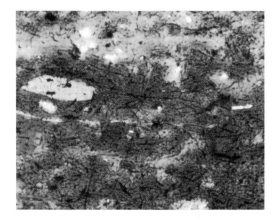

Top: In plane polarized white light a sample of peat dated at about 4000 years old shows cellular structure. Yellow, orange, and red in the middle of the sample are characteristic of wood and peat; the surrounding grayish area is indicative of coaly material already transformed from original vegetable matter. *Bottom:* A coal sample viewed in transmitted light. Red and orange areas are remains of the woody portion of plant material. Yellow areas are other plant components such as spores, leaf cuticles, and resin bodies.

coals and lignites to the high-rank, low-volatile bituminous coals and anthracites. As coal chemists, geologists, and geochemists studied the chemical and crystal structure of coals of various ranks, they developed a general model on which they could agree: as coal was buried deeper in basins, it reached higher temperatures, which forced a partial devolatilization, or loss of gaseous decomposition products, of the coal, bringing it to higher rank. The changes in coal constitution with depth could be measured by the chemistry of solvent extracts of the coal, by X-ray crystallographic methods, by infrared absorption spectra, and by a somewhat unusual technique called optical reflectance.

The reflectance, or the amount of incident light that would be reflected, had long been known to increase with rank. Coal miners knew that the higher ranks shone with a brighter luster. The miners also knew that ordinary bituminous coal was composed of brighter and duller bands. By 1930 coal geologists and paleobotanists had turned to microscopy to pinpoint the botanical origins of the different minerallike entities making up the brighter and duller bands; they recognized the remains of woody tissue, resins and waxes, epidermal materials such as cutin, the tough outer covering of spore, pollen grains, leaves, and the leaflike organs of plants; and they identified some high-carbon fragments as the charcoal resulting from fires in the coal swamp forests. Of all these components, one abundant constituent, vitrinite, stood out for its brightness, its relatively uniform and homogenous character, and its cellular structure, which in many fragments showed its descent from woody tissue. Vitrinite was also the component that showed most unambiguously a regular increase of reflectance with increasing rank.

By 1950, coal researchers had established reflectance as a valuable tool for studying the progression from peat to anthracite. They were beginning to match the reflectance scale with a temperature scale. Anthracitic coal beds interbedded with layers of metamorphosed sedimentary rocks gave the same temperatures as temperatures determined from the temperature-dependent mineral assemblages of the slates and schists. On another front, physical chemists showed that most of the chemical structural changes that moved coal toward an ultimate graphitic structure were dependent only on temperature. For practical purposes, pressure was irrelevant.

Vitrinite reflectance was too good a tool to be the exclusive property of coal geologists. Oil geologists, too, could find little bits and fragments of coaly material commonly incorporated in many sandstones and shales, just as twigs, leaf fragments, and other plant parts are incorporated today in river sands. Near the shoreline, these land-derived materials are still present in small amounts. Only far out to sea do land-plant remains finally disappear. Because these plant fragments are so light, they are more abundant in the slow, quiet, settling environment of muds than in the more energetic sites of sand deposition, where we find only the larger twigs and mats. If we could separate pieces of vitrinite from the sedimentary rock and measure their reflectance, we would have a temperature scale for sedimentary rocks. Such a scale would be particularly applicable to the shale source rocks but less so to reservoir sandstones.

With all these methods for determining the vitrinite reflectance of shales and sandstones, we can easily get temperature estimates of deeply buried rocks—and of the contained oil. However, reflectance is not yet the most precise of tools. The fragments separable from rocks are small and not always easily identifiable as vitrinite, as opposed to other carbonaceous materials that conform to a different reflectance scale. Not surprisingly, as more measurements of reflectance are reported, more ambiguities pop up. Yet it is proving to be one of the powerful methods for taking the temperature of sedimentary rocks.

Oil geochemists were not idle during all these developments. They were well aware that the character of the crude changes with depth of burial. Newer organic analytical methods were used to erect a level of organic maturity (LOM) parameter for oils. Just as coal rank is a measure of the loss of volatile, lower-molecular-weight compounds during the transformation to a high-carbon framework, LOM was tied to the breaking of long-chain, high-molecular-weight hydrocarbons during the transformation to short-chain, low-molecular-weight products such as propane and butane. This scale, too, was largely a function of temperature. The LOM scale joined vitrinite reflectance as a partner in hindcasting past temperatures—LOM for the oil, in one case, and vitrinite for the source rocks or reservoir, in the other. Differences among the temperatures could reveal the history of the oil and sediment.

Correlation between vitrinite reflectance and percentage of carbon, two measures of coal rank, or degree of metamorphism. As coals become higher in carbon content they move from bituminous coal to anthracite.

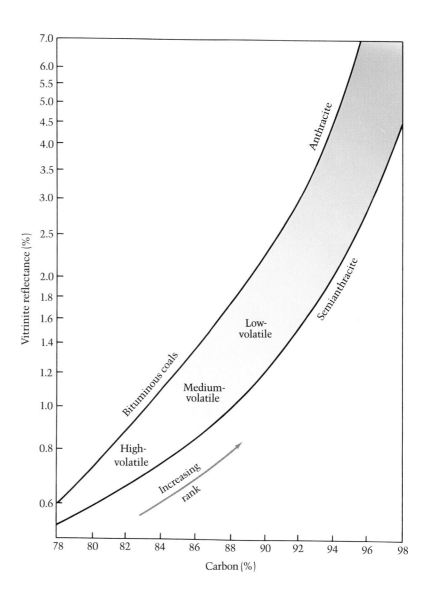

A higher temperature for the oil than for the reservoir might mean that the oil was derived from a deeper, hotter source before migrating to the reservoir. When we interpret temperatures, we need to recognize that oil maturation may precede, accompany, or follow migration. The temperatures themselves led to the idea of an "oil window," or temperature range over which oil would be produced from organic matter: too cool, and oil would never be

Correlation between vitrinite reflectance and the atomic ratios hydrogen/carbon and oxygen/carbon as measures of the increase in maturity of organic matter in oil and gas.

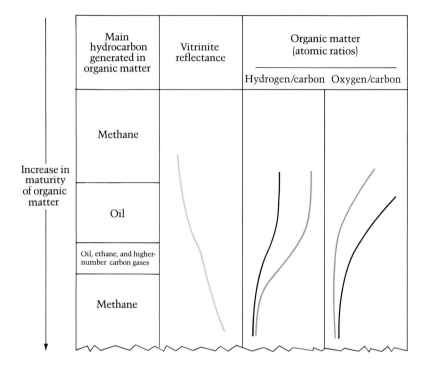

made; too hot, and it would be so cracked that all would be converted to methane, ethane, and propane gases. At even higher temperatures, the lightest gases probably would not be held for long in traps; they would escape through microfractures and the deformational structures associated with metamorphism.

Time plays a role too. All the organic reactions by which oil and gas are produced have slow rates at low temperatures. An oil might be produced over long times at low temperatures as well as over short times at higher temperatures. The current approach is to use a time–temperature integral, or time–temperature index, scaled to the ranges of times and temperatures associated with burial in sedimentary basins. These integrals make clear why prospects for success may diminish as the drill explores deeper, older rocks below younger rocks with mature oil. Because cementation—and the resulting porosity loss—in sandstones also follows some kind of time–temperature relationship, we can use this relationship to predict success or failure of deeper drilling:

The Athabasca tar sands of Alberta, Canada, are mined from an open pit. Conveyer belts along the side carry the sand to a preparation plant where tar is separated from the sand.

our chances to produce the oil or gas may be limited if the sandstone becomes increasingly impermeable with depth.

However, shallower reservoirs are not necessarily better for oil exploration. If an oil-saturated sand remains trapped while being oxidized near the surface, it may go through the same kind of evolution as the tarry residues from oil seeps. The great deposits of the Athabasca tar sands of Canada are hydrocarbons that have thickened to a pitchlike consistency that makes these oils impossible to produce in ordinary ways. These tarry sands might be better mined than pumped. The high viscosity that makes these tars so hard to exploit is the same property that has preserved them from dissipation and loss. Experiments are underway to lower the viscosity of the tar by pumping steam down input wells.

The history of sandstones is tied to the history of the organic matter in them. To understand the maturation oil in sandstone, we need to determine the original organic matter, the rate and depth of burial, the time–temperature history of the oil and the sediment, and the diagenetic history of the sandstone. With all these contingencies, we might expect oil reservoirs to be a rare occurrence and exploration to be difficult. Exploration over the past century has proved this view wrong. Oil is common and ordinary in sedimentary basins, which are themselves common and ordinary formations. If new oil is getting harder and harder to find, it is because we have already found and burned up so much in our profligate demands for energy, not because there is so little or because we do not know how to find it.

CHAPTER TEN

Sandstones evolve

Geologists have to operate with two models of change. According to the dominant, uniformitarian view, past geologic processes were much the same as those we see working today. The less popular model accepts the uniformitarian premise, with the understanding that rates of change may have varied and that conditions and processes—including life—evolved on a planet that developed to maturity through a series of stages. During those stages, especially the earliest, the mechanics of sand formation and preservation may have produced products far different from those of today.

The rock record can inform us only of the last 80 percent of geologic time. We can envision the first billion years only after making calculations based on whichever theory for the origin of the solar system and the planets we consider most plausible. Imagining the second billion years is not quite so bad. In a few places on Earth sedimentary rocks and metamorphic rocks of sedimentary origin of that early age are preserved. Igneous rocks of that period fill in the story of the state of the earliest continents and oceans and the dynamics of crust and mantle that ultimately control surface processes. By the time of the third billion years, we can reconstruct much of the regime, although we are still hampered, as for earlier times, by the lack of a shelly fossil record. Only for the last 600 million years—the Phanerozoic, or "known," part of geologic time—do we have sufficient evidence from both rocks and fossils to be confident of our story. Even in the Phanerozoic, we know far more about the last 100 million years than about earlier periods.

One of the many questions that pique our curiosity about earlier times is how land weathering took place and sand was produced. We know that organisms—from primitive bacteria and

Cross-bedded Navajo sandstone with succulent plant growing in a weathered crevice.

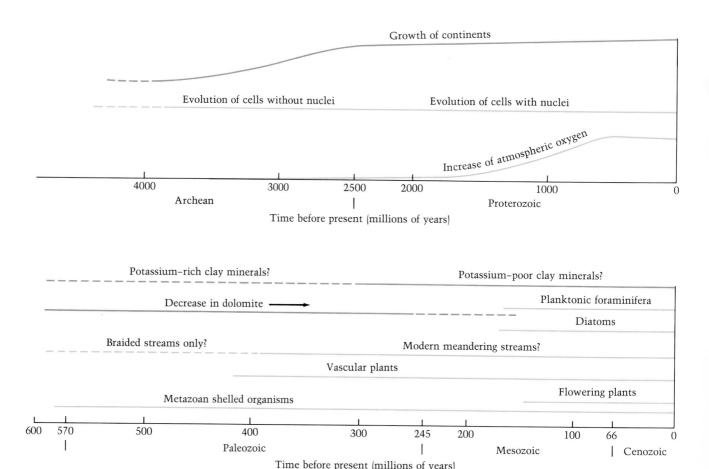

Growth of continents

Evolution of cells without nuclei   Evolution of cells with nuclei

Increase of atmospheric oxygen

| 4000 | 3000 | 2500 | 2000 | 1000 | 0 |

Archean   Proterozoic

Time before present (millions of years)

Potassium–rich clay minerals?   Potassium–poor clay minerals?

Decrease in dolomite ⟶   Planktonic foraminifera

Diatoms

Braided streams only?   Modern meandering streams?

Vascular plants

Flowering plants

Metazoan shelled organisms

| 600 | 570 | 500 | 400 | 300 | 245 | 200 | 100 | 66 | 0 |

Paleozoic   Mesozoic   Cenozoic

Time before present (millions of years)

Some changes in geologic processes and organic evolution that have modified or changed the ways in which sand was produced, transported, and deposited during geologic time *(above)* and during the Phanerozoic, the last 600 million years *(below)*.

fungi to flowering plants—strongly affect the chemistry of rock decay. How did granite weather before the vascular plants, those with cellular differentiation and a body-fluid-transport system, evolved? Before the trees, shrubs, and ground covers first colonized the land surface in the mid-Paleozoic, somewhere around 400 million years ago, the bacteria, algae, and fungi had it all to themselves. Did sand form by the weathering of igneous rocks by the same chemical mechanisms before the appearance of varied greens of the Earth's vegetation. And, much earlier, what was the land surface like before the unicellular organisms existed?

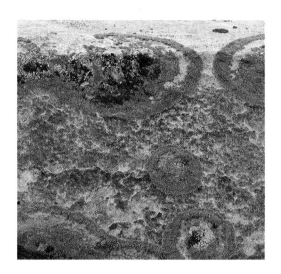

Lichens on sandstone, Canyonlands National Park, Utah. Long before the higher plants evolved, this kind of vegetation covered rocks and contributed to weathering.

Because questions like these take much more knowledge than we now have, geologists constantly compare their deductions about the unknown past with today's regime. For example, an atmosphere without oxygen during the Earth's early periods can be imagined through a favorite thought experiment: let all the present oxygen in the atmosphere magically disappear. How would the Earth system act then? How would rocks reflect such a different regime? One way to approach the many changes in Earth's system through geologic time is to reconstruct the story from the beginning.

TIMES BEFORE THE ROCK RECORD

Our beginning will be the time when this solid planet had already accreted, or gradually grown from the condensed debris of a solar system organizing itself into planets. The earliest sands of a planet cratered and bombarded by planetesimals large and small would have been heterogeneously sized fragments thrown up by violent, explosive impacts. From some combination of our pictures of the Moon and Mars, we can envision a planet large enough to hold an atmosphere, with windblown sands covering its barren surface. As soon as outgassing of the interior provided enormous quantities of water which must have been on the surface from very early times, rains began to pour down and rivers to flow. Sands were transported to the early ocean, there to be spread around on beaches, bars, and other nearshore environments.

What would these beaches have been like? If we were to remove living things from our own beaches, the strand itself would be little altered—except for such obviously missing forms as shells, marine plants, and the bodies of fish thrown up on the sand. Following this thought experiment, we can reason out the nature of the beach environment over 4 billion years ago. Tidal flats beyond the beach would have had no sulfate-reducing bacteria to produce hydrogen sulfide and the black ferrous sulfide. Reddish brown iron oxides would not have been the order of the day, however, because the atmosphere was devoid of plants and therefore of photosynthesis-produced oxygen; it was probably dominated by carbon dioxide and nitrogen oxides, with lesser amounts of reduced gases. The gray sand grains would still have been washed up by the waves, drifted along the beach by longshore currents, and blown back into dunes in back of the beach.

With no land plants to anchor them at their landward edge, the dunes would have been much more extensive, and windblown sand and dust therefore may have been far more widespread over the lands than today. Even temperate regions—assuming climatic zones something like today's—would have had swirling clouds of dust and numerous sandstorms because the winds would easily pick up particles dried soon after the last rain. The strange-looking land surface would have been ruled by erosion and transport of debris, with no roots to hold the soil, no color to enliven the drab, weathered rock.

If we have read correctly the origin of the planet, its atmosphere, and the structure of its early crust, that crust would have been largely igneous. Perhaps it was mostly volcanic, for magmas not far below the surface would have been able to work their way quickly through a thin, hot crust to extrude lava flows and ash falls. These rocks were darker-colored materials, richer in magnesium and iron than the abundant, lighter-colored, silica-rich granitic rocks of more recent stages in the development of continents. The darker minerals and rocks—pyroxenes and amphiboles—are much more unstable than the lighter feldspars during the course of weathering in an oxygen-rich atmosphere. In the early anoxygenic atmosphere, there would have been less difference in stability: all these igneous minerals would have been affected by large amounts of carbon dioxide, which accelerates chemical weathering. Balancing the promotion of weathering by carbon dioxide and the retardation by the absence of oxygen and decay-enhancing organisms, it is likely that weathering and erosion in mountainous regions would have been more rapid than they are today but that weathering in lowlands, which depends so much today on vegetation, would have been less severe. Overall, the ratio of mechanical erosion to chemical weathering might not have been far from today's.

Returning to our primordial beach, we can now imagine the formation of volcanic sands similar to the black beaches of Hawaii and other volcanic islands and offshore sands, largely the product of physical forces, similar to today's. We cannot ignore one possibility: tidal differences then and now. Theories about the origin of this planet include assumptions about the origin of its moon. The Moon's origin by collision of the Earth with a huge, Mars-size planet is currently a favorite theory; others include the tearing of the Moon out of the Earth and the Earth's capture of a passing planet. Whichever account of the early Earth–Moon system you prefer, they all put the Moon much closer to the Earth than it is today, a circumstance that implies a much stronger tidal force and tidal range. Changes in tidal ranges in the open ocean would not have been so important as those in the shallower waters on continental shelves, where tidal flats form.

A final complexity in this game is deducing the existence of continental shelves themselves. The broad expanses of shallow shelves of today are the product of a well-organized plate-tectonic system that creates passive margins of continents as a continent splits along a rift valley, which then spreads to an ocean (see Chapter 7). It is at least as likely in early times that plate tectonics was not the dominant mode of Earth dynamics and that upwelling at volcanic centers formed protocontinents. Such centers would have formed steep declivities to the deeper waters of the ocean, just as modern volcanic islands do. Thus, the tectonic settings hospitable to large, long beaches and extensive tidal flats would not yet have evolved. Our beach, then, would have been the typical small pocket beach of a volcanic island, dispersed intermittently by lava flows or explosive eruptions but reconstituting itself in quieter times.

THE ARCHEAN:
THE EARLIEST RECORD

In posing this example of an early environment as a thought experiment I have not played entirely fair, because we do know a great deal about the Archean, the earlier era of the Precambrian, which spanned the time from the earliest rocks, now dated at about 3.9 billion years, to 2.5 billion years ago. The rocks of the Archean show many elements of the earlier time; the major change was the introduction of life. Direct evidence of microscopic fossils tells us that cellular life began over 3.5 billion years ago, and it could have been earlier. Once the bacteria and algae had evolved, the surface environments of the Earth soon took on appearances much more like those of similar environments of today. Archean sediments contain fine-grain pyrite, which indi-

Filamentous algae of the type that formed stromatolites (algal growth structures in calcium carbonate) in the late Precambrian.

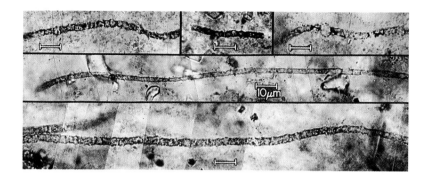

cates that sulfate reducers were active in the Archean. If tidal flats were as organism-rich as they are today, their sands would have looked like today's. We can infer soil-forming bacteria as well, although we cannot be sure of their efficiency relative to the soil-forming work carried out by many sorts of organisms today.

From the sandstones and other sedimentary rocks of the Archean, we cannot tell that life had a significant effect on weathering and sand production. Sandstones and volcanic rocks reflect smaller continents that lacked large continental shelves and whose granitic cores were still small. The main sediments were piles of volcanic rocks and graywacke sandstones, highly feldspathic, dark, mineral-rich rocks, many of which were turbidites, which we defined in Chapter 6. The amounts of quartz-rich sandstones of this age are small, and we have to choose between two explanations for that: either the source rocks were uniformly quartz-poor, or weathering processes that formed quartzose sandstones were not yet in place. The evidence of the igneous rocks dictates which account we pick. Although we find some quartz-rich granites and other quartzose igneous rocks, they are in the minority; instead, quartz-poor igneous rocks such as basalt dominated the scene, both on the continents and in the oceans. Our conclusion therefore follows from both accounts: source rocks were poor in quartz, and weathering processes were not particularly weak.

During this period, areas of continental shelves and shallow-marine waters were small. Calcium-carbonate deposits in the Archean reflect an ocean already at or close to saturation. Some of these carbonate rocks show the structures characteristic of stro-

Stromatolites 1.8 billion years old, Great Slave Lake, Canada.

matolites, algal mats and knobby growths constructed of carbonate particles trapped in the filaments and mucilaginous surfaces of colonial organisms. Associated with these carbonate deposits are sandstones that have all the earmarks of modern shallow-marine sands but one: they have no shells of marine organisms.

THE PROTEROZOIC:
BEGINNINGS OF MODERNITY

The Archean has no well-defined end. The 2.5-billion-year beginning of the next era, the Proterozoic, is a more or less arbitrary time boundary that separates a world of a much more modern aspect from the older times. Large masses of granitic rocks are found on extensive continents, and wide, stable continental shelves, those shallow aprons of sediment built up at passive margins, start to appear. Paleomagnetic measurements of apparent polar wandering, the shift in magnetic pole positions relative to a drifting continent, suggest that the continents at this time started drifting in response to plate motions. Studies of ancient mountain belts show both sedimentary sequences and structural deformation that are consistent with plate tectonics. The full range of sandstone compositions and facies and other sedimentary rock types such as limestones and shales now appear in something like their modern relative abundances. These sediment types were formed by processes like those of today. More-recognizable unicellular life forms, including photosynthetic bacteria, appear as fossils. The atmosphere may not have had very much oxygen, but these photosynthesizers were steadily producing amounts that would later climb to present levels.

The sedimentologists working on Proterozoic sequences might find a regime little different from today's, except for the absence of shelly fossils. For example, the literature is filled with reports of alluvial fining-upward cycles, beach and shallow-marine deposits, sand dunes, and turbidites in the Proterozoic. Sandstone types included all mixtures of source-rock types and ranged from pure quartz sands to highly feldspathic arkoses.

Some differences do appear. Quantitatively and qualitatively, the most special rock type of the Precambrian is the banded iron formation (BIF). Known from both Archean and Proterozoic rocks but absent from the rocks of all later eras, this rock type has provoked more special theories about Precambrian geochemistry than any other type. Banded iron formations, the world's major resource for iron ores, have a distinctive appearance. Thin, alternating bands of quartz and almost pure hematite, or ferric oxide, give a horizontally striped rock. The bands are bedding and represent either alternating conditions of iron oxide precipitation and pure silica deposits or some diagenetic segregation by a chemical process that somehow separated the constituents of an originally mixed rock. We have no recourse to modern environments in which this kind of sediment is forming, for there are none. Although we see some small iron ore deposits of later geologic eras, they have an altogether different character and shed relatively little light on BIFs. Some sandstones associated with BIFs suggest an origin on shallow shelves, and recent geochemical studies indicate that hot springs on or at the base of such shelves may have provided much of the iron and silica of the BIFs. The oxidization we find in most of the beds shows a low level of oxygen, but the level need not be high—it takes very low levels of oxygen to produce hematite from reduced iron. A convincing explanation of BIF sedimentation that satisfactorily accounts for the important properties of composition and distribution still eludes us.

Precambrian banded iron formation, Lake Superior region.

Another strange feature of some Proterozoic sequences is the thickness—several kilometers in some cases—of numerous pure quartz sandstones. Found in the Proterozoic of all continents, quartz sands are common in many parts of North America, from the Uinta Mountains of Utah to eastern Ontario and Quebec. Pure quartz sands are common in later periods, too, but are normally only a few meters to 100 m thick. The great bulk of the Proterozoic sandstones suggests an enormous amount of weathering that eliminated all feldspars, pyroxenes, and other unstable minerals. The lack of any significant amount of interbedded shale also indicates an environment in which all fine-grain particles, silts and muds, were winnowed away. The cross-bedding in the sandstones may indicate shallow-marine sedimentation, but a definitive analysis has yet to be made. These great series of pure quartz sandstones speak of a different time, when abundant quantities of quartz-bearing source rocks had arrived on the continental platforms and weathering processes had ratcheted up in speed, perhaps helped by a colonization of the land surface by newly evolved life forms. The record of huge thicknesses in shallow-water environments speaks of long periods of stability on the young continental shelves.

Other oddities occur in the Proterozoic rock record, but they all tell the story of an adolescent Earth in transition from its infancy to its maturity in the Phanerozoic stage. For us, the vital aspect of the Proterozoic is the staging of life forms that produced differen-

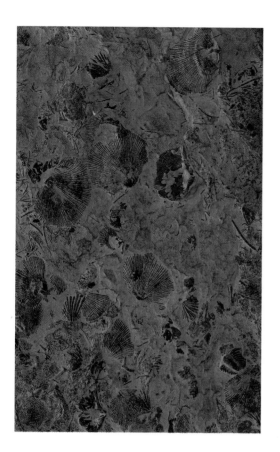

Left: A limestone with abundant fossil shells, typical of the Cambrian and later periods. *Right:* Reconstruction of a swampy forest of the upper Carboniferous period, when many coal beds were formed by the accumulation and burial of this kind of abundant vegetation.

tiated cells, photosynthesis, and, finally, close to the beginning of the Phanerozoic, metazoans, multicellular animals with cells differentiated into tissues and organs. Those differentiated organisms led the way to the further evolution of the shelled forms, the trilobites, corals, molluscs, and all the other organisms whose fossils have dated the later history of this planet.

THE PHANEROZOIC: MATURITY AND CRISES

By the beginning of the Cambrian, 570 million years ago, all the elements of the modern geologic system were in operation. Although some speedups and slowdowns may have occurred, most of the rock record reflects ordinariness from 570 million years ago

to today. During this long time—perhaps short, compared with the 4 billion years that preceded it—the pattern has been slow change punctuated by near-catastrophic episodes.

The evolution of higher organisms is part of the picture. Shelled organisms came first, bequeathing limestones and calcareous sandstones and shales, whose principal textural component is shell material. At the beginning of the Cambrian came the first of the silica-secreting organisms—the single-celled radiolaria and the multicellular sponges—contributing their rock form, chert, made largely of pure silica. In the mid-Paleozoic, the vascular plants evolved on the continents, with significant consequences, we think, for land weathering and landscape. We touched on the weathering story earlier in this chapter and in Chapter 2. Another suggested consequence of vascular land plants is an alteration of

Confluence of two braided streams, Manas, Brazil. Braided patterns such as this may have been the rule before vascular plants arose.

river-channel patterns. As we saw in Chapter 5, rivers today fall into two broad classes, meandering and braided. Which pattern a particular river selects depends on the stream gradient, water and sediment discharge, and climate. One important element determined by climate is the vegetation on the banks, which resists the river's erosion of the banks and encourages the river to meander. It may be that the braided river was the norm before this binding vegetation evolved.

The rise and fall of various invertebrate shelled organisms helped to determine the kinds of limestones deposited at different times but had little influence on sandstone deposition. The evolution of flowering plants in the Cretaceous period also had little effect on sandstones. The dinosaurs came and went with no impact on sand sedimentation, and so did the mammals—until the genus *Homo* arrived on the scene a short time ago. For most of prehistory and history, we barely put a dent in our physical surroundings, but in the last century we have dammed and channelized rivers and irrigated deserts to provide more arable land. Many

Aerial view of the Colorado River delta at the head of the Gulf of California. The river is at flood stage and the tide is high. For much of the year so much water is removed from the river for public water supplies and irrigation that it runs dry at its mouth and is unable to transport sand.

arid lands have become deserts through overgrazing and overpopulation, and the carbon dioxide we have pumped into the atmosphere may warm the Earth sufficiently to melt some of the polar ice caps. Rates of erosion on all the continents have immeasureably speeded up as a result of the machinery of modern agriculture—and greater erosion has meant increased rates of sedimentation of sands, muds, and all the other sediments that derive from land erosion.

Beaches are a good example of this process. Hardly a stretch of sandy shoreline in economically developed countries has been exempt from our interference. We engineer breakwaters, groins, and barriers to prevent sand from eroding, and we usually make matters worse. The latest idea is to vacuum clean the sand to remove paper cartons, plastics of all descriptions, and beer cans. (Sedimentologists have yet to count these materials as normal parts of a sand body.) The needs of construction have destroyed the fragile balance of erosion, sedimentation, and vegetation of

A sandy soil area cleared for housing development in northern San Diego. Some of these kinds of development can affect local rates of erosion and groundwater conditions and, in hilly areas, may accelerate landslides or mudslides, but probably do not affect the global pattern of sand production and sedimentation.

many of the world's barrier islands. To be fair, we have not only destroyed beaches but have sometimes constructed them in front of hotels that never had them before.

We could look at our activity as the work of yet another organism modifying its environment—just as the lowly clam makes a home in the sand by burrowing—but on a grander scale. Still, our work, like that of smaller organisms, is on too small a scale to change the dynamics of sand transport and deposition. With all the engineering and construction, we hardly make a difference in the total amount of beach sand over the globe. Our works, good or bad, are local and temporary. An abandoned beach quickly reverts to its former steady state—and the future looks like the past. Or does it?

SANDS OF THE FUTURE

In *The Time Machine*, H. G. Wells forecast a time well into the future.

> The machine was standing on a sloping beach. The sea stretched away to the south-west, to rise into a sharp bright horizon against the wan sky. There were no break-

ers and no waves, for not a breath of wind was stirring. Only a slight oily swell rose and fell like a gentle breathing, and showed that the eternal sea was still moving and living. And along the margin where the water sometimes broke was a thick incrustation of salt—pink under the lurid sky. . . . I noticed that I was breathing very fast . . . from that I judged the air to be more rarefied than it is now.

Fifty million years further came the red giant stage of the sun: "the huge red-hot dome of the sun had come to obscure nearly a tenth part of the darkling heavens . . . the red beach . . . seemed lifeless."

Predictions of the long-term future of the Earth have long been the property of science fiction writers. Earth scientists are currently finding it difficult enough to estimate the changes in the atmosphere and oceans over the next 50 years that will result from carbon dioxide increases. Nevertheless, geologists occasionally wonder out loud how the world will look some millions or billions of years from now. Until the mid-1960s, we had some outlines of stellar evolution to guide our thoughts on the evolution of the sun and our less-formed ideas on entropy in the Earth's interior as its heat engine runs down. Since the acceptance of plate-tectonic theory, we have a more definite dynamic view of the future Earth's geography. We can expect the Atlantic Ocean to continue spreading and expanding and the Pacific to shrink as its lithosphere sinks into the subduction zones that border it. If past geological history is a guide, the two oceans may switch roles, with the Atlantic becoming ringed with volcanic fire as the continents around it converge.

Some hundreds to thousands of million years hence, the plate-tectonic machine will run down as heat supplied from radioactivity in the crust and mantle is dissipated. Heat loss by conduction and convection will cool the interior, and mantle convection will slow. As a consequence of convection slowing, plates will move at a more leisurely pace, the lithosphere under both continents and oceans will thicken, hot spots will cool, and the dynamic Earth will start creaking to a halt. Tectonically, the Earth will freeze at whatever state it happens to have reached.

When tectonics ceases, erosion and sedimentation become the dominant forces of change. All mountains will be gradually low-

Cross section of the crust and upper mantle of the Earth as it might look many hundreds of millions or billions of years in the future when the planet has cooled to the point where plate tectonics no longer will operate.

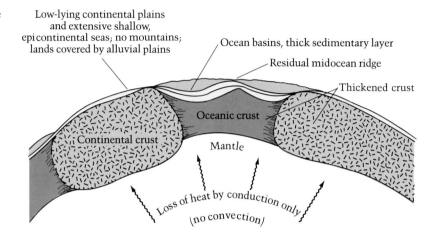

ered to low hills by erosion while the sediment produced by that erosion blankets the low-lying continents and continental shelves. The broad expanses of continental shelf will surround all continents as tectonically active continental margins, those tied to subduction and volcanic arcs, disappear, and passive continental margins, like those of the present Atlantic Coast of North America, become the exclusive mode of continent–ocean borders. Sand will be distributed over the continental and shallow-water marine environments, as they are today, and will become increasingly quartz-rich as continued weathering of the lowlands dissolves the feldspars and other unstable minerals. With tectonics over and the Earth in a transition period, the subsidence of sedimentary basins will slow to a halt; cooling and contraction of all formerly hot regions of the lithosphere will bring all provinces to the uniformly low thermal gradients of stability.

Sand will no longer be produced when all continental surfaces lie only a few tens of meters above the ocean. Except during floods, the sand that lay over alluvial plains, deltas, and continental borderlands will be covered by muds. Unrejuvenated and weathering deeply, the soils of the world will be impoverished in nutrients but rich in bauxite, the oxide–hydroxide ore of aluminum, and in iron and titanium oxides. How all this would affect the Earth's climate, its atmosphere, and oceans, is incalculable at the moment, even by general predictions. The effects on Earth's life we will leave to our imaginative guides to the future, science fiction writers.

We read a cloudy crystal ball about sand in a posttectonic world because sand is the product of tectonism, of the dynamics of Earth's interior interacting with the atmosphere and waters of the globe's surface. The sand we dribble through our fingers reminds us of mountains and rivers, deserts and beaches, volcanoes and glaciers. Thus, we return to Blake and the universe in a grain of sand. For most of us, the universe is our experience of the land. To know sand is to know that land.

Further reading

CHAPTER ONE

Press, F., and R. Siever, 1986, *Earth*, 4th ed. New York, W. H. Freeman and Company, for a general geological background.
Pettijohn, F. J., P. E. Potter, and R. Siever, 1987, *Sand and Sandstone*, 2d edition, New York, Springer-Verlag, for a more advanced treatment of the geology of sand.
Leeder, M. R., 1982, *Sedimentology*, Winchester, Mass., Allen and Unwin, gives an introduction to sedimentology.

CHAPTER TWO

Loughman, F. C., 1969, *Chemical Weathering of the Silicate Minerals*, New York, Elsevier.
Krauskopf, K., 1979, *Introduction to Geochemistry* (Chapter 4), 2d ed., New York, McGraw-Hill.
Millot, G., Clay, *Scientific American*, April 1979 (offprint 937).
Blatt, H., G. Middleton, and R. Murray, 1980, *Origin of Sedimentary Rocks* (Chapter 7), 2d ed., Englewood Cliffs, N.J., Prentice-Hall.

CHAPTER THREE

Kuenen, P. H., 1960, Sand, *Scientific American*, April 1960.
Allen, J. R. L., 1985, *Principles of Physical Sedimentology*, Winchester, Mass., Allen and Unwin, is especially good for fluid flow in relation to sediment transport and sedimentary structures.
Middleton, G. V., and J. B. Southard, 1977, *Mechanics of Sediment Movement, Short Course 3*, Tulsa, Okla., Society of Economic Paleontologists and Mineralogists.
Potter, P. E., and F. J. Pettijohn, 1977, *Paleocurrents and Basin Analysis*, 2d ed., New York, Springer-Verlag.

CHAPTER FOUR

Walker, R. G. (Ed.), 1984, *Facies Models,* 2d ed., Toronto, Geoscience Canada/Geological Association of Canada.

Reading, H. G., (Ed.), 1986, *Sedimentary Environments and Facies,* 2d ed., New York, Elsevier.

Schumm, S. A., 1977, *The Fluvial System,* New York, Wiley.

Leopold, L. B., and W. B. Langbein, River Meanders, *Scientific American,* June, 1966.

Glennie, K. W., 1970, *Desert Sedimentary Environments,* New York, Elsevier.

Bagnold, R. A., 1941, *The Physics of Blown Sand and Desert Dunes,* London, Methuen.

CHAPTER FIVE

Davis, R. A., Jr., 1978, *Coastal Sedimentary Environments,* New York, Springer-Verlag.

Komar, P. D., 1976, *Beach Processes and Sedimentation,* Englewood Cliffs, N.J., Prentice-Hall.

Ginsburg, R. N. (Ed.), *Tidal Deposits,* New York, Springer-Verlag.

Broussard, M. L. (Ed.), 1975, *Delta Models for Exploration,* Houston, Houston Geological Society.

Strahler, A. N., 1966, *A Geologist's View of Cape Cod,* New York, American Museum of Natural History.

CHAPTER SIX

Bouma, A. H., W. R. Normark, and N. E. Barnes (Eds.), 1985, *Submarine Fans and Related Turbidite Systems: Frontiers in Sedimentary Geology.,* Vol. 1, New York, Springer-Verlag.

Kuenen, P. H., and C. I. Migliorini, 1950, Turbidity Currents as a Cause of Graded Bedding, *Journal of Geology* 58:91–117.

Stanley, D. J., and G. Kelling (Eds.), 1978, *Sedimentation in Submarine Canyons, Fans, and Trenches,* Stroudsburg, Penn., Dowden Hutchinson and Ross.

CHAPTER SEVEN

Dewey, J. F., Plate Tectonics, *Scientific American,* May 1972.

Hallam, A., 1973, *A Revolution in the Earth Sciences: From Continental Drift to Plate Tectonics,* New York, Elsevier.

Takeuchi, S., S. Uyeda, and H. Kanamori, 1970, *Debate about the Earth* (Rev. Ed.), San Francisco, Freeman-Cooper.

Courtillot, V., and G. E. Vinks, How Continents Break Up, *Scientific American*, July 1983.

Dickinson, W. R., 1974, Plate tectonics and sedimentation, in W. R. Dickinson (Ed.), *Tectonics and Sedimentation,* Tulsa, Okla., Society of Economic Paleontologists and Mineralogists, Spec. Pub. 22, pp. 1–27.

CHAPTER EIGHT

Scholle, P. A., and P. R. Schluger (Eds.), *Aspects of Diagensis,* Tulsa, Okla., Society of Economic Paleontologists and Mineralogists Spec. Pub. 26.

Taylor, J. C. M., 1978, *Sandstone Diagenesis,* Journal of the Geological Society of London Pub. 135, pp. 6–135.

Parker, A., and B. W. Sellwood (Eds.), 1983, *Sediment Diagenesis* (NATO Advanced Study Institute Series), Dordrecht–Boston–Lancaster, D. Reidel.

Berner, R. A., 1980, *Early Diagenesis,* Princeton, N.J., Princeton University Press.

CHAPTER NINE

Tissot, B. P., and D. H. Welte, 1978, *Petroleum Formation and Occurrence,* New York, Springer-Verlag.

Hunt, J. M., 1979, *Petroleum Geochemistry and Geology,* San Francisco, W. H. Freeman and Company.

Levorsen, A. I., 1967, *Geology of Petroleum,* 2d ed., San Francisco, W. H. Freeman and Company.

CHAPTER TEN

Stanley, S. M., 1985, *Earth and Life Through Time,* New York, W. H. Freeman and Company.

Holland, H. D., 1984, *The Chemical Evolution of the Atmosphere and Oceans,* Princeton, N.J., Princeton University Press.

Windey, B. F., 1976, *The Early History of the Earth* (NATO Advanced Study Institute Proceedings), New York, Wiley.

Sources of illustrations

page 54
William Garnett

page 60
Larry Ulrich

page 62
Adapted from P. E. Potter, 1955, Petrology and Origin of the Lafayette Gravel, *Journal of Geology 63*

page 63
Adapted from F. Press and R. Siever, 1986, *Earth*, 4th ed., W. H. Freeman and Company, New York

page 64
Top, adapted from A. V. Jopling, 1963, Hydraulic Studies on the Origin of Bedding, *Sedimentology 2*; bottom, adapted from J. R. L. Allen, 1968, *Current Ripples*, North Holland, Amsterdam

page 65 (flume)
A. V. Jopling

page 66
Heather Angel/Biofotos

page 67
Adapted from J. C. Harms, J. B. Southard, and G. V. Middleton, 1982, *Structures and Sequences in Clastic Rocks, Short Course 9*, Tulsa, Okla Society of Economic Paleontologists and Mineralogists

CHAPTER FOUR
page 68
Bill Ratcliffe

page 70 (fluvial environment)
Adapted from R. G. Walker, and D. J. Cant, 1984, Sandy Fluvial Systems, in R. G. Walker (Ed.), *Facies Models*, 2d ed., Geoscience Canada/Geological Association of Canada, Toronto

page 71 (fining upward)
Adapted from F. Press and R. Siever, 1986, *Earth*, 4th ed., W. H. Freeman and Company, New York

page 73
left, Tom Bean; right, William Garnett

page 75
GEOPIC™, Earth Satellite Corporation

page 79
Adapted from L. B. Leopold and W. B. Langbein, 1966, River Meanders, *Scientific American*, June

page 80
Heather Angel/Biofotos

page 81
Adapted from J. R. L. Allen and P. F. Friend, Deposition of the Catskill Facies, Appalachian Region, with Notes on Some Other Old Red Sandstone Basins, 1968, in G. Klein (Ed.), *Late Paleozoic and Mesozoic Continental Sedimentation, Northeastern North America*, Geological Society of America Special Paper 106

page 82
William Garnett

page 83
Bill Ratcliffe

page 84
Adapted from R. A. Bagnold, 1941, *The Physics of Blown Sand and Desert Dunes*, Methuen, London

page 85
left and center, Tui de Roy; right, Bill Ratcliffe

page 86
Betty Crowell/Faraway Places

page 87
Tom Bean

page 88
Tom Bean

CHAPTER FIVE
page 90
GEOPIC™, Earth Satellite Corporation

page 92
From The Floor of the Oceans, based on bathymetric studies by B. C. Heezen and M. Tharp, copyright © Marie Tharp, 1977

page 95
From C. C. Bates, 1953, *Rational Theory of Delta Formation*, American Association of Petroleum Geologists Bulletin 37

page 96
right, GEOPIC™, Earth Satellite Corporation

page 97
Adapted from H. E. Reineck, 1970, Marine sandkörper, rezent und fossil, *Geol. Rundschau 60*

page 98
top, adapted from R. M. Mitchum et al., 1977, American Association of Petroleum Geologists Memorandum 26; bottom, adapted from S. M. Stanley, 1986, *Earth and Life Through Time*, W. H. Freeman and Company, New York

page 99
William Garnett

page 100
Adapted from L. D. Wright and J. M. Coleman, 1973, American Association of Petroleum Geologists Bulletin 57

page 101
Comprised from various sources, including C. R. Kolb and J. R. Van Lopik 1966, in M. L. Shirley (Ed.), *Deltas in Their Geological Framework*, Houston Geological Society; and D. E. Frazier, 1967, Recent Deltaic Deposits of the Mississippi River: Their Development and Chronology, Gulf Coast Association of the Geological Society Transactions 17

page 103
Adapted from F. Press and R. Siever, 1986, *Earth*, 4th ed., W. H. Freeman and Company, New York

page 179
Scott Blackman/Tom Stack and
Associates

page 180
Holger Jannasch, Woods Hole
Oceanographic Institute

page 182
Adapted from F. Press and R. Siever,
1986, *Earth,* 4th ed., W. H. Freeman
and Company, New York

page 184
Gerald and Lysbeth Corsi/Focus on
Nature

page 185
William Garnett

page 190
Adapted from R. Tada and R. Siever,
1986, Experimental Knife-edge
Pressure Solution of Halite,
*Geochimica et Cosmoschimica
Acta 50,* © 1986 by Permagon Press
Ltd.

page 191
R. Tada

CHAPTER NINE

page 192
B. F. Molnia/Terra Photographics-
BPS

page 195
Betty Crowell/Faraway Places

pages 198–201
From A. I. Levorsen, 1967, *Geology
of Petroleum,* 2d. ed., W. H.
Freeman and Company, New York

page 203
U.S. Bureau of Mines

page 204
Adapted from H. Blatt, *Sedimentary
Petrology,* 1982, W. H. Freeman and
Company, New York

page 205
From B. P. Tissot and D. H. Welte,
1978, *Petroleum Formation and
Occurrence,* Springer-Verlag, New
York

page 207
American Petroleum Institute and
Exxon Corporation

CHAPTER TEN

page 208
Philip Hyde

page 210
Adapted from F. J. Pettijohn, P. E.
Potter, and R. Siever, 1987 *Sand
and Sandstone,* 2d ed., Springer-
Verlag

page 211
Philip Hyde

page 214
J. W. Schopf

page 215
S. M. Awramik/BPS

page 217
Travis Amos

page 218
Bill Ratcliffe

pages 218–219
left, Bill Ratcliffe; right, courtesy,
Field Museum of Natural History
[trans. no. 75400], Chicago

page 220
GEOPIC™, Earth Satellite
Corporation

page 221
Peter Kresan

page 222
Bruce Molnia

Index

Other Books in the Scientific American Library Series